Modeling and Analysis
of Compositional Data

STATISTICS IN PRACTICE

Series Advisors

Human and Biological Sciences
Stephen Senn
CRP-Santé, Luxembourg

Earth and Environmental Sciences
Marian Scott
University of Glasgow, UK

Industry, Commerce and Finance
Wolfgang Jank
University of Maryland, USA

Founding Editor
Vic Barnett
Nottingham Trent University, UK

Statistics in Practice is an important international series of texts which provide detailed coverage of statistical concepts, methods and worked case studies in specific fields of investigation and study.

With sound motivation and many worked practical examples, the books show in down-to-earth terms how to select and use an appropriate range of statistical techniques in a particular practical field within each title's special topic area.

The books provide statistical support for professionals and research workers across a range of employment fields and research environments. Subject areas covered include medicine and pharmaceutics; industry, finance and commerce; public services; the earth and environmental sciences, and so on.

The books also provide support to students studying statistical courses applied to the above areas. The demand for graduates to be equipped for the work environment has led to such courses becoming increasingly prevalent at universities and colleges.

It is our aim to present judiciously chosen and well-written workbooks to meet everyday practical needs. Feedback of views from readers will be most valuable to monitor the success of this aim.

A complete list of titles in this series appears at the end of the volume.

Modeling and Analysis of Compositional Data

Vera Pawlowsky-Glahn

University of Girona, Spain

Juan José Egozcue

Technical University of Catalonia, Spain

Raimon Tolosana-Delgado

Helmholtz Institut Freiberg for Ressources Technology, Germany

Library of Congress Cataloging-in-Publication Data

Pawlowsky-Glahn, Vera.
  Modelling and analysis of compositional data / Vera Pawlowsky-Glahn, Juan José Egozcue, Raimon Tolosana-Delgado.
       pages cm
  Includes bibliographical references and indexes.
  ISBN 978-1-118-44306-4 (cloth)
  1. Multivariate analysis. 2. Mathematical statistics. 3. Geometric analysis. I. Egozcue, Juan José, 1950-
II. Tolosana-Delgado, Raimon. III. Title.
  QA278.P39 2015
  519.5′35–dc23

                                                                                                  2014043243

A catalogue record for this book is available from the British Library.

ISBN: 9781118443064

Set in 10.5/12.5pt, Times-Roman by Laserwords Private Limited, Chennai, India

1   2015

We cannot solve our problems
with the same thinking
we used when we created them.

Albert Einstein

Eppur si muove

Galileo Galilei

Contents

Preface

This book is an illustration of the adage collected by Thomas Fuller in *Gnomologia* (1732, Adage 560): *All things are difficult, before they are easy* and cited by John Aitchison (1986, Chapter 3). It has been a long way to arrive at this point, and there is still a long and not always easy way to go in the light of the insights presented here. Therefore, we dedicate this work to all those researchers who are not mainstream and have to struggle swimming against the tide.

These pages are based on lecture notes originally prepared as support to a short course on compositional data analysis. The first version of the notes dates back to the year 2000. Their aim was to transmit the basic concepts and skills for simple applications, thus setting the premises for more advanced projects. The notes were updated over the years, reflecting the evolution of our knowledge about the geometry of the sample space of compositional data. The recognition of the role of the sample space and its algebraic-geometric structure has been essential in this process. This book reflects the state of the art at the beginning of the year 2014. Its aim is still to introduce the reader into the basic concepts underlying compositional data analysis, but it goes far beyond an introductory text, as it includes advanced geometrical and statistical modeling. One should also be aware that the theory presented here is a field of active research. Therefore, the learning process can just start with this book, and a study of the latest contributions presented at meetings and as articles in journals is strongly recommended.

The book relies heavily on the monograph *"The Statistical Analysis of Compositional Data"* by John Aitchison (1986) and on posterior fundamental developments that complement the theory developed there, mainly those by Aitchison (1997), Barceló-Vidal et al. (2001), Billheimer et al. (2001), Pawlowsky-Glahn and Egozcue (2001, 2002), Aitchison et al. (2002), Egozcue et al. (2003), Pawlowsky-Glahn (2003), Egozcue and Pawlowsky-Glahn (2005), and Mateu-Figueras et al. (2011). Specific literature for other aspects of compositional analysis is given in the corresponding chapters. Chapter 1 gives a brief overview of the history of these developments and presents some everyday examples to illustrate the need of compositional data analysis. Chapter 2 defines compositions and their characteristics and introduces their sample space, the simplex. Zeros and other irregular components are addressed in Section 2.3. On the basis of these considerations, Chapter 3 presents the Aitchison geometry of the simplex, while Chapter 4 gathers several ways to represent compositional data within this geometry. These four chapters form the algebraic-geometric body of the book, the backbone of the rest of the material.

Chapter 5 deals with exploratory analysis techniques adapted to compositions. Chapter 6 covers some distribution models for random compositions, as well as some required elements of probability theory. In particular, the latter chapter includes the normal distribution on the simplex, essential for the following two chapters. They are devoted to advanced statistical modeling: Chapter 7 provides some tools for testing compositional hypotheses (numerically and graphically), while Chapter 8 focuses on linear models, including regression, analysis of variance, and discriminant analysis. The last two chapters give an overview of what lies beyond this book: Chapter 9 outlines several compositional models besides the linear model, while the epilogue (Chapter 10) summarizes the ongoing and open aspects of research, as well as further topics, too specific to deserve longer attention in a general-purpose book.

Readers should take into account that, for a thorough understanding of compositional data analysis, a good knowledge in standard univariate statistics, basic linear algebra, and calculus, complemented with an introduction to applied multivariate statistical analysis, is a must. The specific subjects of interest in multivariate statistics, developed under the assumptions that the sample space is the real space with the usual Euclidean geometry, can be learned in parallel from standard textbooks, for instance, Krzanowski (1988) and Krzanowski and Marriott (1994) (in English), Fahrmeir and Hamerle (1984) (in German), or Peña (2002) (in Spanish). Thus, the intended audience goes from advanced students in applied sciences to practitioners, although the original lecture notes proved to be useful for statisticians and mathematicians as well. Newcomers to the field may find specially useful to start with Chapters 1–3, then read the first five sections of Chapter 4 and switch to Chapters 5 and 7 before finishing up Chapter 4. Applied practitioners already familiar with the basics of compositional data analysis should have a look at the notation and concepts in Chapters 4 and 6, before passing to the modeling Chapters 7–9. This book includes an extensive list of references, two appendices with practical recipes and some basic elements of random variables, a list of the symbols used in the book, and two indices; an author index and a general index. In the latter, pages in boldface indicate the point where the corresponding concept is defined.

Concerning notation, it is important to note that, to conform to the standard praxis of registering multivariate observations as a matrix where each row is an observation or data point and each column is a variate, vectors will be considered as row vectors (denoted by square brackets) to make the transfer from theoretical concepts to practical computations easier. Furthermore, as a general rule, theoretical parameters will be denoted by either Latin or Greek letters and their estimators by the same letters with a hat.

Throughout the book, examples are introduced to illustrate the concepts presented. The end of each example is indicated with a diamond suit ($\Diamond$).

Most chapters end with a list of exercises. They are formulated in such a way that many can be solved using an appropriate software. CoDaPack is a user friendly, cross-platform, freeware to facilitate this task, which can be downloaded from the web. Details about this package can be found in Thió-Henestrosa and Martín-Fernández (2005) or Thió-Henestrosa et al. (2005). Those interested in working with R (or S-plus) may use the packages "compositions" by Boogaart and Tolosana-Delgado (2005, 2013) in general or "robCompositions" by Templ et al. (2011) for robust compositional data analysis, as well as their common graphical user interface "compositionsGUI" by Eichler et al. (2013).

Vera Pawlowsky-Glahn
Juan José Egozcue
Raimon Tolosana-Delgado

About the Authors

 **Dr. Vera Pawlowsky-Glahn** is professor at the University of Girona, Department of Computer Science, Applied Mathematics, and Statistics. She studied Mathematics at the University of Barcelona (UB), Spain, and obtained her PhD (doctor rerum naturam) from the Free University of Berlin, Germany. Before going to Girona, she was professor in the School of Civil Engineering at the Technical University of Catalonia (UPC) in Barcelona. Her main research topic since 1982 has been the statistical analysis of compositional data. The results obtained over the years have been published in multiple articles, proceedings, and books. Together with A. Buccianti she has acted as editor of a book in honor of J. Aitchison in 2011 published by Wiley, who will also publish in 2015 a textbook on modeling and analysis of compositional data, co-authored with J. J. Egozcue and R. Tolosana-Delgado. She was the leader of a research group on this topic involving professors from different Spanish universities. The group organizes every two years a workshop on compositional data analysis, known as CoDaWork, and their research has received regularly financial support from the Spanish Ministry for Education and Science and from the University Department of the Catalan Government. Prof. Dr. Pawlowsky-Glahn has been vice-chancellor at UPC from 1990 to 1994, head of the Department of Computer Science and Applied Mathematics at the University of Girona in 2004–2005, and dean of the Graduate School of the University of Girona in 2005–2006. She received in 2006 the William Christian Krumbein Medal of IAMG, was nominated Distinguished Lecturer of IAMG in 2007, and received the J.C. Griffiths Teaching Award in 2008. From 2008 to 2012 she was President of IAMG and is now Past-President.

Dr. Juan José Egozcue studied Physics, oriented to Geophysics and Meteorology, at the University of Barcelona (Spain). He obtained his PhD in Physics in the same university with a dissertation on maximum entropy spectral analysis (1982). In 1978 he got a position as a lecturer in the school of civil engineering in Barcelona (Escuela de Ingeniería de Caminos, Canales y Puertos de la Universidad Politécnica de Cataluña (UPC), Barcelona, Spain), teaching several topics on Applied Mathematics. In 1983 he started teaching Probability and Statistics. He became Full Professor in 1989, at the UPC, where he has been Vice-Chancellor of the university (1986–1988) and Chair of the Department of Applied Mathematics III (1992–1998).

His research activities are presently centered in two lines: estimation of natural hazards using Bayesian methods, specially applied to seismic, rainfall and ocean wave hazards; and analysis of compositional data, with special emphasis in the geometry of the sample space.

He started research on compositional data analysis around 2000 in cooperation with Dr. Vera Pawlowsky-Glahn. The first results appeared in 2001–2002 when the Euclidean vector space structure of the simplex recognized. The development of this geometry led to the introduction of the isometric logratio transformation for compositional data and the concept of balance (2003–2005) which have proven their usefulness in a number of applied fields.

Dr. Raimon Tolosana-Delgado completed in 2002 a degree in Engineering Geology in Barcelona, at the School of Civil Engineers (UPC) and the Faculty of Geology (UB), and in 2004 a Master in Environmental Technology and Physics at UdG, all along focusing on compositional data analysis and Geostatistical methods applied to Earth Sciences. In 2006 he completed his PhD under the supervision of Dr. Pawlowsky-Glahn, on the spatial analysis of data from restricted spaces, as a generalization of compositional data analysis. Since then, he has been working as a fellow researcher between Spain and Germany, working with compositional models in sedimentology at the University of Göttingen, and later back at the UPC, in weather and climate modeling and data assimilation through geostatistical simulations and restricted space consideration. Since October 2012 he has been applying and developing compositional and spatial methods as a researcher at the Department of Modeling and Valuation, Helmholtz Institute Freiberg for Resource Technology in Freiberg (Saxony, Germany), a joint research institute of the Technical University "Bergakademie" Freiberg and the Helmholtz Zentrum Dresden-Rossendorf dealing with all aspects of the value chain of Rare Earths and other technological elements, from mineral exploration and mining to technological waste recycling.

Acknowledgments

We acknowledge the many comments made by readers of the original lecture notes, pointing at both small and important errors in the text. Essential have also been the many contributions and discussions presented at several editions of CoDaWork, the International Workshop on Compositional Data Analysis, extensively cited throughout the text. They all have contributed to improve the theory presented here. We also appreciate the support received from our institutions, research groups, the *Spanish Ministry of Economy and Competitiveness* under the project "METRICS" (Ref. MTM2012-33236), and the Generalitat de Catalunya through the project "Compositional and Spatial Analysis" (COSDA) (Ref. 2014SGR551).

1

Introduction

Compositional data describe parts of some whole. They are commonly presented as vectors of proportions, percentages, concentrations, or frequencies. As proportions are expressed as real numbers, one is tempted to interpret, or even analyze, them as real multivariate data. This practice can lead to paradoxes and/or misinterpretations, some of them well known even a century ago, but mostly forgotten and neglected over the years. Some simple examples illustrate the anomalous behavior of proportions when analyzed without taking into account the special characteristics of compositional data.

Example 1.1 (Intervals covering negative proportions).
Daily measurements of an air pollutant are reported as $3 \pm 5 \ \mu g/m^3$. The given interval of concentration covers a nonsensical range of concentrations that includes negative values. It is probably generated by an average of concentrations which contain some values much higher than $3 \ \mu g/m^3$. For instance, the following is a set of rounded random percentages: $1, 1, 2, 3, 4, 4, 7, 13, 29, 37$. Their mean is 10.1%, while their standard deviation is 12.7%. Thus a typical $2s$-interval for the mean value would be an interval covering negative proportions, namely, $(-15.3\%; 35.5\%)$. A frequent procedure is to cut this interval at zero, but then the question arises on what happens to the probability assigned to the eliminated part of the interval, $(-15.3\%; 0\%)$, and to the probability assigned to the retained part, $(0\%; 35.5\%)$. ◇

Modeling and Analysis of Compositional Data, First Edition.
Vera Pawlowsky-Glahn, Juan José Egozcue and Raimon Tolosana-Delgado.
© 2015 John Wiley & Sons, Ltd. Published 2015 by John Wiley & Sons, Ltd.

Example 1.2 (Small proportions: Are they important?).
Frequently, when some components or parts of a composition are very small, they are eliminated, with the argument that they are negligible. In such a case, it is important to think about *the salt in a soup*. Consider a soup that is perfectly seasoned to your taste, and imagine somebody adds to the soup the same amount of salt you used, thinking that it was not yet seasoned. Probably, doubling the amount of salt will spoil it completely. To our understanding, this is a perfect example on how important a small proportion can be and why a relative scale gives you better information in this case than an absolute one. Sometimes, small proportions are added to other parts, for example, salt and other spices, but that leads to a loss of information, making the recipe insufficiently specified.   ◇

Example 1.3 (Reporting changes in proportions).
In the 1998 election to the German Bundestag, the German Liberal Party (FDP) obtained 6.2% of the votes. Eleven years later, in the 2009 elections, they obtained a share of 14.6%. This could be reported as an increment of 8.4 percentage points. We are more used to reading that FDP increased its proportion of votes a 135% ($6.2 + 6.2 \times 135/100 = 14.6$). In the following election, just 4 years later, the party decreased its votes by a significant 67%, but still half of the increment that occurred between 1998 and 2009. Nevertheless, that meant that the FDP was not anymore represented in the Bundestag, because its share ($14.6 - 14.6 \times 67/100 = 4.8$) dropped below the threshold of 5% required by the German electoral law. How can it be that increasing 135% and decreasing 67% gives a negative balance? Perhaps this is a bad way of reporting changes in proportions (data extracted from Wikipedia (2014)).

Reporting increments of shares in differences of percentage points have also disappointing properties, as the relative scale of proportions is ignored. In fact, an increment of 8.4 percentage points represents a very important change from the 1998 result of FDP (6.2%). It would be not so important if the previous 1998 result were, for instance, 30%.   ◇

Example 1.4 (The scale of proportions).
In a given year, the annual proportion of rainy days in a desert region is 0.1%, and near a mountain range it is 20.0%. Some years later, these proportions have changed to 0.2% and 20.1%, respectively. To summarize the situation, one can assert that the rainy days in both regions have increased by 0.1%. Such a statement suggests the idea of a homogeneous change in the two different regions, ignoring that the rainy days in the desert have been doubled, while in the mountain range the proportion is almost the same. Using the increment of ratios typical of election results or economic reports, the rainy days would have increased a critical 100% in the desert, and a slightly relevant 5% in the mountains.

Furthermore, if some analysis of the evolution of the rainy days is made in both regions, it should be guaranteed that equivalent results are obtained if the nonrainy days are analyzed. In the desert region, the annual proportion of non-rainy days has changed from 99.9% to 99.8% and near the mountain range from 80.0% to 79.9%. That represents that nonrainy days have decreased, respectively, 0.001% and 0.00125%, which suggests almost no difference between the mountain and the desert. How can it then be that rainy days change so dramatically in the desert and nonrainy days do not change at all? A proper analysis should assure that no paradoxical results are obtained when analyzing one type of days and its complementary.   ◇

Example 1.5 (The Simpson's paradox).
The lectures on statistics started very early this morning. Students (men and women) are divided into two classrooms. Some of them arrived on time and some of them were late. Academia was interested in knowing about punctuality according to the gender of the students. Therefore, data were collected this morning during the statistics lectures. The data set is reported in Table 1.1. The paradoxical result is that, for both classrooms, the proportion of women arriving on time is greater than that of men. On the contrary, if the individuals of both classrooms are joined in a single population, the proportion of punctual men is larger than that of the women. This kind of paradoxical results are known as Simpson's paradox (Simpson, 1951; Julious and Mullee, 1994; Zee Ma, 2009). The paradox can be viewed from different points of view. The simplest one, the arithmetic perspective, is to look at the way in which proportions are aggregated: to find the proportion of on-time women in the joint population, the per class-room proportions a_1/W_1, a_2/W_2 are *averaged* as $(a_1 + a_2)/(W_1 + W_2)$, where a_i is the number of on-time women in the classroom i and W_i is the corresponding

Table 1.1   Number of students of two classrooms, arriving on time and being late, classified by gender. Proportions are reported under the number of students. The largest proportion of arriving on-time men and women are in boldface for easy comparison.

	Classroom 1		Classroom 2		Total	
	On time	Late	On time	Late	On time	Late
Men	53	9	12	6	65	15
	0.855	0.145	0.667	0.333	**0.813**	0.188
Women	20	2	50	18	70	20
	0.909	0.091	**0.735**	0.265	0.778	0.222

total of women. This kind of average is ill-behaved for proportions as shown by Simpson's paradox.

A second point of view is to look at the total proportion of on-time women as a mean value of this proportion in the two classrooms. Each classroom is treated as a sample individual and $(a_1 + a_2)/(W_1 + W_2)$ is taken as the sample mean of the proportions. The paradoxical result suggests that mean values of proportions should be redefined carefully to get consistent results.   ◊

Example 1.6 (Spurious correlation).
The Spanish Government publishes the number of affiliations to the Social Security on a monthly basis, which is classified into the following categories depending on the type of company: agricultural, industrial, construction, and service. The 144 data, corresponding to a monthly series going from 1997 to 2008, were downloaded from the corresponding web site (Gobierno de España, 2014). A version, prepared for processing, is available in (www.wiley.com/go/glahn /practical). First, to obtain proportions between the different types of company, the data were normalized to add to 1 in the full composition comprising the four categories. Then, the correlation matrix was computed (see Table 1.2). Next, to analyze the behavior of the companies excluding *construction*, a subcomposition of three categories was obtained, suppressing the category *construction* and converting the three-part vector to proportions, so that the three components add up to 1. Again, the correlation matrix was computed (see Table 1.3). When analyzing correlations in the full composition with four parts and the subcomposition with three parts, the correlation between the proportion of agricultural and industrial companies only changed slightly, actually from -0.9808 to -0.9887, whereas the correlation between the service companies and either agricultural or industrial companies changed dramatically, from 0.1699 to 0.9863 in the first case and from -0.0723 to -0.9999 in the second. This is a typical effect when analyzing a set of parts adding up to a constant, or a subset of the same parts, closed to any constant.

Table 1.2   Correlation of proportion of affiliations to social security in Spain according to the type of company (four-part composition: agricultural, industrial, construction, and service).

	Agricultural	Industrial	Construction	Service
Agricultural	1.0000	−0.9808	0.9201	0.1699
Industrial	−0.9808	1.0000	−0.9663	−0.0723
Construction	0.9201	−0.9663	1.0000	−0.1867
Service	0.1699	−0.0723	−0.1867	1.0000

Table 1.3   Correlation of proportion of affiliations to social security in Spain according to the type of company (three-part subcomposition: agricultural, industrial, and service).

	Agricultural	Industrial	Service
Agricultural	1.0000	−0.9887	0.9863
Industrial	−0.9887	1.0000	−0.9999
Service	0.9863	−0.9999	1.0000

The problem of spurious correlation is sometimes circumvented by avoiding the closure when considering a subcomposition. This is equivalent to say: the percentages of agricultural, industrial, construction, and service affiliates constitute a composition as the percentages add to 100%; to overcome the compositional intricacies, we can remove one component, for example, service, so that the remaining percentages do not add to 100%. This way, the correlation matrix between the percentages of agricultural, industrial, and construction affiliates are exactly those reported in Table 1.2 in the first three columns and rows. However, a new question arises: what would happen if we start with two additional categories of affiliation closed to 100%?   ◇

The awareness of problems related to the statistical analysis of compositional data dates back to a paper by Karl Pearson (1897) the title of which began significantly with the words *"On a form of spurious correlation ... "*. Since then, as stated in Aitchison and Egozcue (2005), the way to deal with this type of data has gone through roughly four phases, which can be summarized as follows:

Phase I: 1897–1960

Karl Pearson, in his paper on spurious correlations, pointed out the problems arising from the use of standard statistical methods with proportions. But his warnings were ignored until around 1960, despite the fact that a compositional vector – with components the parts of some whole – is usually subject to a constant-sum constraint.

Phase II: 1960–1980

Around 1960, the geologist Felix Chayes (1960) took up the problem and warned against the application of standard multivariate analysis to compositional data. He tried to separate what he called the *real* from the *spurious* correlation, in an attempt to avoid the *closure problem*, expressed mainly as a negative bias induced by the constant-sum constraint. Important contributions in geological applications were made, among others, by Sarmanov and Vistelius (1959), and Mosimann (1962) which drew the attention of biologists. However, as pointed out by Aitchison and Egozcue (2005), *distortion of standard multivariate techniques when applied to compositional data was the main goal of study.*

Phase III: 1980–2000

Aitchison, in the 1980s, realized that compositions provide information about relative, not absolute, values of parts or components. Consequently, every statement about a composition can be stated in terms of ratios of components (Aitchison, 1981, 1982, 1983, 1984). The facts that logratios are easier to handle mathematically than ratios, and that a logratio transformation provides a one-to-one mapping onto a real space, led to the advocacy of a methodology based on a variety of logratio transformations. These transformations allowed the use of standard unconstrained multivariate statistics applied to transformed data, with inferences translatable back into compositional statements. But they were, not without difficulties, derived from the fact that the usual Euclidean geometry and measure were implicitly assumed for the sample space of compositional data.

This phase deserves special attention because transform techniques have been very popular and successful over more than a century; from the Galton-McAlister introduction of the logarithmic transformation for positive data, through variance-stabilizing transformations for sound analysis of variance, to the general Box–Cox transformation (Box and Cox, 1964) and the implied transformations in generalized linear modeling. The logratio transformation principle is based on the fact that there is a one-to-one correspondence between compositional vectors and associated logratio vectors, so that any statement about compositions can be reformulated in terms of logratios, and vice versa. The advantage is that the problem of a constrained sample space, the simplex, is removed. Data are projected into multivariate real space, opening up all available standard multivariate techniques. The original transformations were principally the additive logratio transformation (Aitchison, 1986, p. 113) and the centered logratio transformation (Aitchison, 1986, p. 79). The logratio transformation methodology seemed to be accepted by the statistical community; see, for example, the discussion of Aitchison (1982).

Phase IV: 2000–present

Around 2000, several scientists realized independently that the internal simplicial operation of perturbation, the external operation of powering, and the simplicial metric define a metric vector space (indeed a Hilbert space) (Billheimer et al., 1997, 2001, Pawlowsky-Glahn and Egozcue, 2001). The recognition of the algebraic-geometric structure of the sample space of compositions led to the staying-in-the-simplex approach for the analysis of compositional problems (Pawlowsky-Glahn, 2003). This approach is essentially based on the *principle of working in coordinates* (Mateu-Figueras et al., 2011). Compositions are represented by orthonormal coordinates, which live in a real Euclidean space. They can be interpreted in themselves or from their representation in the simplex. The sample space of random compositions is represented by the simplex with a simplicial metric and measure, different from the usual Euclidean metric and Lebesgue measure in real space.

The book presented here corresponds to the fourth phase. It summarizes the state of the art in the stay-in-the-simplex approach. Therefore, the first part will be devoted to the algebraic-geometric structure of the simplex, which we call *Aitchison geometry* (Pawlowsky-Glahn and Egozcue, 2001). Although it is a Euclidean geometry, we felt that it is important to distinguish it from the usual geometry in real space.

2

Compositional data and their sample space

2.1 Basic concepts

Let us start with the definition of the central concept in this book, namely, the type of data the methods presented here have been developed for.

Definition 2.1 (*D*-part composition).
A (row) vector, $\mathbf{x} = [x_1, x_2, \ldots, x_D]$, *is a D-part composition when all its components are strictly positive real numbers and carry only* relative information.

The meaning of *relative information* refers to that where the only information is contained in the ratios between the components of the composition and the numerical value of each component by itself is irrelevant. Indeed, that compositional information is relative is implicitly stated in the units, as they are usually parts of a whole. The most common examples have *a constant sum* κ and are known in the literature as *closed data* (Chayes, 1971). Frequently, $\kappa = 1$, which means that measurements have been made in, or transformed to, parts per unit (proportions), or $\kappa = 100$, for measurements represented in percentages. Other units are also possible, such as ppm (parts per million) or ppb (parts per billion), which are typical examples for compositional data where only a part of the composition has been recorded; or, as some studies show, even concentration

Modeling and Analysis of Compositional Data, First Edition.
Vera Pawlowsky-Glahn, Juan José Egozcue and Raimon Tolosana-Delgado.
© 2015 John Wiley & Sons, Ltd. Published 2015 by John Wiley & Sons, Ltd.

units (mg/L, meq/L, molarities, and molalities), where no constant sum can be feasibly defined (Buccianti and Pawlowsky-Glahn, 2005; Otero et al., 2005). In these cases, a straightforward transformation, such as division (multiplication) by the molar weight, allows the transformation to (from) proportions. Therefore, the term *compositional data* includes all types of data representing parts of some whole, in whichever units this occurs.

The fact that multiplication of a vector of positive components by a positive constant does not change the ratios between the components suggests that compositions can be viewed as equivalence classes made of proportional vectors, all of them conveying the same compositional information. This is reflected in the following definition.

Definition 2.2 *(Compositions as equivalence classes).*
Two vectors of D positive real components $\mathbf{x}, \mathbf{y} \in \mathbb{R}_+^D$ (x_i, $y_i > 0$, for all $i = 1, 2, \ldots, D$) are compositionally equivalent if there exists a positive constant $\lambda \in \mathbb{R}_+$ such that $\mathbf{x} = \lambda \cdot \mathbf{y}$.

To change the units of the data (for instance, from % to ppm), each component is multiplied by the constant of change of units (from % to ppm by 10^4). Both vectors, the initial in % and the resulting one in ppm, have their components proportional. They are compositionally equivalent or, more properly, they represent the same composition. From a geometric point of view, a vector of D strictly positive components can be represented as a point P' in the positive orthant of $\mathbb{R}^D$ (for three components see Figure 2.1a). The half-line or ray connecting the origin with P' is made of other, proportional, vectors. They are representatives of an equivalence class, that is, a composition.

Any point of an equivalence class can be used to represent it. A traditional and practical possibility is to select constant sum vectors as representatives of the compositions. In other words, any composition can be expressed in proportions using the appropriate scaling factor. In Figure 2.1a, the constant sum representative of P' is P, which lies on the plane containing the vectors that components sum up to a given constant. The operation of assigning a constant sum representative to a composition is called *closure* and is defined as follows.

Definition 2.3 *(Closure).*
For any vector of D strictly positive real components,

$$\mathbf{z} = [z_1, z_2, \ldots, z_D] \in \mathbb{R}_+^D, \qquad z_i > 0 \qquad for\ all \qquad i = 1, 2, \ldots, D,$$

the closure of $\mathbf{z}$ to $\kappa > 0$ is defined as

$$C(\mathbf{z}) = \left[\frac{\kappa \cdot z_1}{\sum_{i=1}^{D} z_i}, \frac{\kappa \cdot z_2}{\sum_{i=1}^{D} z_i}, \ldots, \frac{\kappa \cdot z_D}{\sum_{i=1}^{D} z_i} \right].$$

The result of closure is a rescaling of the initial vector so that the sum of its components is κ. The definition of compositionally equivalent vectors can be rephrased using closure: two vectors $\mathbf{x}$, $\mathbf{y}$ in $\mathbb{R}^D_+$ are compositionally equivalent if $C(\mathbf{x}) = C(\mathbf{y})$, whichever is the constant κ.

Consequently, compositional data can be represented as proportions using an appropriate scaling factor. For ease of mathematical operations involved, from now on, we will assume compositional data represented as proportions, that is, as vectors of constant sum κ. The set of possible compositions (see Definition 6.5 for the associated statistical concept of *sample space*) can then be determined.

Definition 2.4 *(Sample space).*
The sample space of compositional data is the simplex,

$$S^D = \left\{ \mathbf{x} = \left[x_1, x_2, \ldots, x_D \right] \middle| x_i > 0,\ i = 1, 2, \ldots, D;\ \sum_{i=1}^{D} x_i = \kappa \right\}. \qquad (2.1)$$

Although, properly speaking, a composition is an equivalence class, the representatives of such classes in the simplex are also called *compositions*. The components of a vector in S^D are called *parts* to remark their compositional character. Note that κ depends on the units of measurement or rescaling: usual values are 1 (proportions), 100 (%), 10^6 (ppm), and 10^9 (ppb).

Frequently, interest is focused on only some parts of a composition. This motivates the introduction of subcompositions.

Definition 2.5 *(Subcomposition).*
Given a composition $\mathbf{x}$ *and a selection of indices* $S = \{ i_1, \ldots, i_s \}$, *a subcomposition* $\mathbf{x}_S$, *with s parts, is obtained by applying the closure operation to the subvector* $[x_{i_1}, x_{i_2}, \ldots, x_{i_s}]$ *of* $\mathbf{x}$. *The set of subscripts S indicate which parts are selected in the subcomposition, not necessarily the first s ones.*

Very often, compositions contain many parts; for example, the major oxide bulk composition of igneous rocks has around 10 elements, and they are but a few of the total possible. In a stock market, many companies participate, and the corresponding stock returns can be viewed as a composition. However, the subcomposition of returns from companies integrated in a given index can be of interest.

Actually, one seldom represents the full composition. In fact, most of the applied literature on compositional data analysis (mainly in geology) restricts their figures to three-part (sub)compositions. For three parts, the simplex can be represented as an equilateral triangle (Figure 2.1a), with vertices at $A = [\kappa, 0, 0]$, $B = [0, \kappa, 0]$ and $C = [0, 0, \kappa]$. It is commonly visualized in the form of a *ternary or de Finetti diagram* – which is an equivalent representation. A ternary

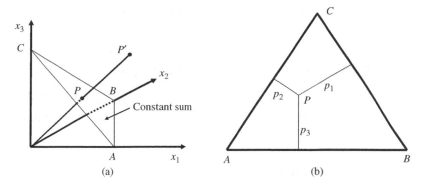

Figure 2.1   (a) Simplex imbedded in $\mathbb{R}^3$. (b) Ternary diagram.

diagram is an equilateral triangle such that a generic sample $\mathbf{p} = [p_1, p_2, p_3]$ will plot at a distance p_1 from the opposite side of vertex A, at a distance p_2 from the opposite side of vertex B, and at a distance p_3 from the opposite side of vertex C (Figure 2.1b). The triplet $[p_1, p_2, p_3]$ is commonly called the *barycentric coordinates* of $\mathbf{p}$, easily interpretable but useless in plotting (plotting them would yield the three-dimensional plot of Figure 2.1a). What is needed to get the plot of Figure 2.1b is the expression of the coordinates of the vertices and the samples in a Cartesian coordinate system $[u, v]$, and this is given in Appendix A.

Finally, if only some parts of the composition are available, a fill-up or residual value can be defined, or simply the observed subcomposition can be closed. Note that because one seldom analyses every possible part, in practice, only subcompositions are used. In any case, both methods (fill-up or closure) should lead to identical or, at least, compatible results.

Apart from building subcompositions, another way commonly found for reducing the dimensionality of a compositional data set is to amalgamate some components, that is, to sum them into one new part.

Definition 2.6 (Amalgamation).
Given a composition $\mathbf{x} \in S^D$, and a selection of a indices $A = \{i_1, \ldots, i_a\}$ (not necessarily the first ones), $D - a \geq 1$, and the set of remaining indices $\bar{A}$, the value

$$x_A = \sum_{j \in A} x_i$$

is called amalgamated part *or* amalgamated component. *The vector $\mathbf{x}' = [\mathbf{x}_{\bar{A}}, x_A]$, containing the components with subscript in $\bar{A}$ grouped in $\mathbf{x}_{\bar{A}}$ and the amalgamated component x_A, is called* amalgamated composition *which is in S^{D-a+1}.*

Note that using a fill-up or residual value is equivalent to using an amalgamated composition.

2.2 Principles of compositional analysis

Three conditions should be fulfilled by any statistical method that is to be applied to compositions: scale invariance, permutation invariance, and subcompositional coherence (Aitchison, 1986).

2.2.1 Scale invariance

The most important characteristic of compositional data is that *they carry only relative information*. Let us explain this concept with a couple of examples.

Example 2.7 (Energy production in the EU).
According to EUROSTAT (2014), the whole EU produced in 2007 roughly 857 Mtonnes of oil equivalent energy from primary sources, split in 481.5 for all hydrocarbon sources, 241.5 for nuclear energy, and 134 for those known as renewable sources. They can be represented as a composition $x_0 = [56.2, 28.2, 15.6]\%$, clearly dominated by hydrocarbon sources. The EU directives established as a target that 20% of energy should come from renewable sources, a situation that was reached in 2011, where the shares of sources were $x_1 = [50.6, 29.2, 20.2]\%$. Without knowing what happened with the total primary energy production, several scenarios can be devised to explain those changes. A first scenario could assume a constant renewable production and an adjustment of the other two sources, specially a reduction of hydrocarbon production: for instance, keeping the renewable production stable, a vector of productions $x_R = [335, 193.5, 134]$ (in Mtonnes), yielding a total production of 662.5 Mton would be compatible with the observed proportions. A second scenario could imply a stable oil production, together with an increase of the other two sources, to $x_H = [481.5, 192.5, 278]$ Mtonnes totaling 952.5 Mtonnes. Did the EU reach its goal of renewable production *share* by massive development of renewable production (as implied by scenario 2) or was it rather an effect of a radical reduction of hydrocarbon production (scenario 1), for instance, because of the crisis-induced reduction of total demand? Actually, just knowing the percentages given by x_1, it cannot be decided which of the two scenarios corresponds to reality.   ◇

Example 2.8 (Erosion and reworking of sands).
In a paper with the suggestive title *Unexpected trend in the compositional maturity of second-cycle sands* (Solano-Acosta and Dutta, 2005), the lithologic composition of a sandstone and its derived recent sands was analyzed looking at

the percentage of grains made up of only quartz (Q), only feldspar (F), or rock fragments (R). For medium-sized grains coming from the parent sandstone, an average composition $[Q, F, R] = [53, 41, 6]\%$ is reported, while for the offspring sands, the mean values reported are $[37, 53, 10]\%$. One expects that feldspar and rock fragments decrease as the sediment matures; thus they should be less important in a second-generation sand. *Unexpectedly* (or apparently so), this does not happen in the example. To pass from the parent sandstone to the offspring sand, several different changes are possible, yielding exactly the same final composition. Assume those values were weight percent (in g/100 g of bulk sediment). Then, one of the following could have happened:

1. Q suffered no change passing from sandstone to sand, but per 100 g parent sandstone 35 gF and 8 gR were added to the sand (for instance, because of comminution of coarser grains of F and R from the sandstone);

2. F was unchanged, but per 100 g parent sandstone 25 gQ were depleted, and at the same time, 2 gR were added (for instance, because Q was better cemented in the sandstone, tending to form coarser grains);

3. any combination of the former two extremes.

The first two cases yield, respectively, per 100 g parent sandstone, final masses of $[53, 76, 14]$ g and $[28, 41, 8]$ g. In a purely compositional data set, we do not know whether mass was added or subtracted from the sandstone to the sand. Thus, which of these cases actually occurred could not be decided. Without further (noncompositional) information, there is no way to distinguish between $[53, 76, 14]$ g and $[28, 41, 8]$ g, as we only have the value of the sand composition *after closure*. Closure (Definition 2.3) is a projection of any point in the positive orthant of D-dimensional real space onto the simplex. As stated in Definition 2.2, all points on a ray starting at the origin (e.g., $[53, 76, 14]$ and $[28, 41, 8]$) are projected onto the same point of S^D (in this case, $[37, 53, 10]\%$). The ray is an *equivalence class* and the point on S^D a *representative* of the class. Figure 2.2 shows this relationship. Moreover, to change the units of the data keeping the sum constant (for instance, from % to ppm), simply multiply all the points by the constant of change of units, moving them along their rays to the intersections with another triangle, parallel to the plotted one.   ◇

In the absence of information about the total (total power production or mass of sediment), it is highly reasonable to expect analyses to yield the same results, in whichever way that total evolved. This is known as *scale invariance* (Aitchison, 1986).

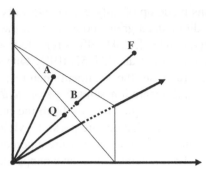

Figure 2.2   Representation of the compositional equivalence relationship.
A *represents the original sandstone composition,* **B** *the final sand composition,*
F *the amount of each part if feldspar was added to the system (first hypothesis),*
and **Q** *the amount of each part if quartz was depleted from the system (second*
hypothesis). Note that the points **B**, **F**, *and* **Q** *are compositionally equivalent.*

Definition 2.9 (*Scale invariance*).
Let $f(\cdot)$ be a function defined on $\mathbb{R}_+^D$. This function is scale invariant if for any
positive real value $\lambda \in \mathbb{R}_+$ and for any composition $\mathbf{x} \in S^D$ it satisfies $f(\lambda\mathbf{x}) =$
$f(\mathbf{x})$, that is, it yields the same result for all compositionally equivalent vectors.

Mathematically speaking, this is achieved if $f(\cdot)$ is a 0-degree homogeneous function of the parts in $\mathbf{x}$. Practical choices of such functions are logratios of the
parts in $\mathbf{x}$ (Aitchison, 1997; Barceló-Vidal et al., 2001). To illustrate why this is
so, let us assume that $\mathbf{x} = [x_1, x_2, \ldots, x_D]$ is a composition given in percentages.
The ratio $f(\mathbf{x}) = x_1/x_2 = (\lambda \cdot x_1)/(\lambda \cdot x_2)$ is scale invariant and yields the same
results if the composition is given in different units, for example, in parts per
unit or in ppm, because units cancel in the ratio. However, ratios are strictly positive and depend on the ordering of parts, because $x_1/x_2 \neq x_2/x_1$. A convenient
transformation of ratios is the corresponding logratio, $f(\mathbf{x}) = \ln(x_1/x_2)$. Now, the
inversion of the ratio only produces a change of sign, thus giving a symmetry to
$f(\cdot)$ with respect to the ordering of parts.

More complicated logratios are useful. For instance, define

$$f(\mathbf{x}) = \ln \frac{x_1^{\alpha_1} \cdot x_2^{\alpha_2} \ldots x_s^{\alpha_s}}{x_{s+1}^{-\alpha_{s+1}} \cdot x_{s+2}^{-\alpha_{s+2}} \ldots x_D^{-\alpha_D}} = \sum_{i=1}^{D} \alpha_i \ln x_i,$$

where powers α_i are real constants (positive or negative). The α_i's are assumed
negative for $i = s+1, s+2, \ldots, D$, thus appearing in the denominator of the
ratio with a positive value $-\alpha_i$. For this logratio to be scale invariant, the sum

of all powers should be null. Scale-invariant logratios are called logcontrasts (Aitchison, 1986).

Definition 2.10 *(logcontrast).*
Consider a composition $\mathbf{x} = [x_1, x_2, \ldots, x_D] \in S^D$ *and coefficients* $\alpha_i \in \mathbb{R}$ *for all* $i = 1, 2, \ldots, D$. *A logcontrast is a function*

$$f(\mathbf{x}) = \sum_{i=1}^{D} \alpha_i \ln x_i, \quad with \quad \sum_{i=1}^{D} \alpha_i = 0.$$

In applications, some logcontrasts are easily interpreted. A typical example is chemical equilibrium. Consider a chemical D-part composition, denoted by $\mathbf{x}$ and expressed in molar proportions. A chemical reaction involving four species may be

$$\alpha_1 x_1 + \alpha_2 x_2 \rightleftharpoons \alpha_3 x_3 + \alpha_4 x_4,$$

where other parts are not involved. The α_i's, called stoichiometric coefficients, are normally known. If the reaction is matter preserving, then $\alpha_1 + \alpha_2 = \alpha_3 + \alpha_4$. The logratio

$$\ln \frac{x_1^{\alpha_1} \cdot x_2^{\alpha_2}}{x_3^{\alpha_3} \cdot x_4^{\alpha_4}}$$

is a logcontrast because $\alpha_1 + \alpha_2 - \alpha_3 - \alpha_4 = 0$. Whenever this chemical reaction is in equilibrium, this logcontrast should be constant and, therefore, deviations from it can be interpreted as departures from equilibrium (Boogaart et al., 2013).

2.2.2 Permutation invariance

A function is *permutation invariant* if it yields equivalent results when the ordering of the parts in the composition is changed. The following two examples illustrate what *equivalent* means here. The *distance* between the initial sandstone and the final sand composition should be the same working with $[Q, F, R]$ or working with $[F, R, Q]$ (or any other *permutation* or reordering of the parts). Also, if interest lies in the *change* occurred from sandstone to sand, results should be equal after reordering. A usual way to get rid of some numerical problems occurring with some classical multivariate statistics is to remove one component: this procedure is not permutation invariant, as results will largely depend on which component is suppressed.

However, ordered compositional data are frequent. A typical case corresponds to the discretization of a continuous variable. Some interval categories are defined on the span of the variable, and then the number of occurrences in each category is recorded as frequencies. These frequencies can be considered as a composition, although categories are still ordered, and the order is important. The information concerning the ordering will not be taken into account in a standard compositional analysis.

2.2.3 Subcompositional coherence

The final condition is *subcompositional coherence*: subcompositions (Definition 2.5) should behave like orthogonal projections in real analysis. The size of a projected segment is less than, or equal to, the size of the segment itself. This general principle, although shortly stated, has several practical implications, explained in the next chapters. The most illustrative, however, are the following.

- The distance between two full compositions must be greater than, or equal to, the distance between them when considering any subcomposition. This particular behavior of the distance is called *subcompositional dominance*. Exercise 4 shows that the Euclidean distance between compositional vectors does not fulfill this condition and is thus ill-suited to measure distance between compositions.

- If a noninformative part is removed, results should not change. For instance, if hydrogeochemical data are available, and interest lies in classifying the kind of rocks washed by the water, in general, the relations between some major oxides and ions will be used (SO_4^{2--}, HCO_3^-, Cl^-, to mention just a few), and the same results should be obtained taking milliequivalents per liter (including implicitly water content), or weight percent of the ions of interest.

Subcompositional coherence can be practically summarized as: (i) distances between two compositions are equal or decrease when subcompositions of the original ones are considered; (ii) scale invariance of the results is preserved within arbitrary subcompositions, that is, the ratios between any parts in the subcomposition are equal to the corresponding ratios in the original composition (Egozcue, 2009).

2.3   Zeros, missing values, and other irregular components

2.3.1   Kinds of irregular components

The principle of scale invariance of Section 2.2.1 led to the conclusion that compositions should always be treated in terms of logratios. It implies that zero components cannot be directly dealt with in this context, as logarithms of zero values are undefined. Unfortunately, zeros occur quite frequently in real-world data sets. Zeros and missing values are thus an important issue of the methodology presented in this book. The present section introduces several concepts of missing or zero values, and some useful strategies to deal with them. These are further developed from an applied perspective in Chapters 5 and 8.

It must be mentioned that the problem of missing values is already difficult in standard multivariate statistical analysis. In compositional analysis, it becomes even more complex, as one single missing component yields the whole composition undefined: if one component is not available, how can the total sum be calculated in order to apply the closure (Definition 2.3)? If it is considered a missing value, the total sum is missing as well, and all components of the closed vector will be missing. If it is considered zero, then the total sum is actually computed with respect to the observed components, and we will be in fact working with their subcomposition, instead of with the original composition. Thus, it is convenient not to apply any closure or normalization to compositions with missing values.

Attending to the nature of the irregular components, the following kinds of problematic elements may appear.

Rounded zeros, censored values, or values below detection limit

Rounded zeros correspond to a value which is reported as zero, although the component is present in a very small quantity, below the number of decimals used for reporting the values. A similar concept is that of a value below the detection limit: in that case, a value is reported as zero because the detection method could not prove its presence, that is, the presence of that component could not be proven to be different from zero (Boogaart et al., 2011). In both cases, it is suggested that the *zero* should be replaced by something *small*, below or around the rounding/detection limit. Examples often occur in chemical analyses. Owing to the typical noise of the measuring machine, any presence below a threshold (called *detection limit*) cannot be analytically distinguished from total absence. The common practice in chemometry is to report it as zero. Note that these are special cases of censored data: with censored data, the true value is missing, but one has instead a surrogate information, such as a credible interval. Censored data are not necessarily always smaller than a bound. For instance, some measuring devices break down when the value is *larger* than a threshold, thus meaning that the true value was above it. A treating strategy in this case should be based on replacing the censored value by a value above the threshold.

Structural zeros

A structural zero is a value that is intrinsically zero because of a physical limitation. Actually, it has no sense to consider that component. When dealing with data sets with some structural zeros, it is usual to consider that the population with zeros is different from the population without zeros (Aitchison, 1986; Bacon-Shone, 2003). Typical examples are the proportion of expenditures in tobacco for a nonsmoking family or of meat consumption in a vegan family; the proportion of feldspatoids in granite (because granite must have loads of quartz, and feldspatoids and quartz cannot crystallize from the same melt); or the votes to a political party in an election district in which it does not present any candidate (a related worked example is given in Section 5.9).

Counting zeros (and counts in general)

A different, but related, kind of zeros may occur when the composition is derived from closing a vector of counts. A zero in this case has something in common with the detection limit (perhaps the count could have been 1 if the sample size had been larger), but it actually points to a deeper problem. A vector of counts may not be a composition in the strict sense of the term, because scale invariance might not apply, as illustrated in the following examples. Consider measuring the popularity of several parties by counting their votes in an election in several districts of very different size (considered as the total number of voters): observing votes to two parties of 20 versus 30 is not fully equivalent to 20,000 versus 30,000 votes, even though they show the same relation 2 : 3. Consequently, a party obtaining 0 votes in the first of the two districts might have had bad luck, but if it obtains 0 votes in the second district, it has a flawed program. This zero is *more fundamental or structural* in the second case, while it is more a *rounding zero* in the first case. Similar examples can be found in petrographic classifications of sands, where a fixed number of grains are classified in certain types (and some types might not be caught, though being present). It is not the same to count 40 grains than to count 400: a zero in the second case suggests a much smaller frequency of a certain mineral than a counting zero in the first case. Another example is encountered when evaluating ecosystems by counting insects of a certain taxon captured by a trap (species that are not captured after using the trap for 1 month could be captured if the trap was left for half a year). All these cases of count compositions might produce counting zeros, but even when such vectors do not show zeros, it may be better to treat them with latent variable models (see below). Note that, strictly speaking, these considerations virtually affect all compositional data, as they are mostly obtained by closing some vectors of counts. To consider them scale invariant, and thus, compositional, is in the end a decision of the analyst.

Missing values

Missing values are those which were not even measured or reported or which got lost for some reason. Several mechanisms can generate missing values:

1. the probability of missing a particular value depends on the missing component itself, that is, values are *not missing at random* (NMAR);

2. the probability of missing a particular value depends only on the observed components, that is, values are *missing at random* (MAR);

3. the probability of missing a particular value does not depend on any of them, that is, the values are *missing completely at random* (MCAR).

Each case has its own possible way of solving the problem, although most of the times one might try to model the possible dependencies and exploit them to devise a treatment process. Examples of missing values are elements that are sometimes not measured because the analytical techniques are too expensive, elements that are accidentally polluted in the laboratory and are not reliable anymore, or questions in a survey that a respondent did not answer.

Amalgamation

One or more variables may be reported as zero or missing value, but their contribution was actually added to another component or components. These components are then said to be *amalgamated*. Amalgamation is incompatible with the techniques presented in this book, and undoing it is extremely problematic. Amalgamation should be either avoided or applied to the whole data set. Examples of amalgamated data are (i) parties that participate in an election sometimes or somewhere in coalition (see Section 5.9); (ii) international educational level surveys, where some countries consider master and PhD levels together and some other consider them separately; (iii) amalgamation of chemical elements, such as Na and K in hydrochemical analyses, a frequent practice.

More about zeros and missing values, their definitions, as well as strategies to deal with them can be found in Aitchison and Kay (2003), Bacon-Shone (2003), Boogaart and Tolosana-Delgado (2013), Fry et al. (2000), Greenacre (2011), Martín-Fernández et al. (2000), Martín-Fernández (2001), Martín-Fernández et al. (2003), Martín-Fernández et al. (2011), and Martín-Fernández and Thió-Henestrosa (2006).

2.3.2   Strategies to analyze irregular data

Replacement

The replacement strategy is the most intuitive one, and the first that appeared in the literature (Aitchison, 1986). The idea is simple: take all those values giving problems and replace each one of them by a likely nonproblematic value. The best example of this procedure is the replacement of a rounded zero below a detection limit by a fixed or a random fraction of the detection limit itself, for example, 1/2 or 2/3 of the detection limit (Martín-Fernández et al., 2003). But the same can be applied to missing values, replaced by an estimated mean (or a random value around it). To undo an amalgamation, two or more components that have been amalgamated can be replaced by splitting the amalgamated quantity among them in a way consistent with the rest of the sample.

A given replacement is sensible if it is compatible with the information we have about that lost value. For example, a below-detection-limit value should still be below the detection limit after the replacement; replaced amalgamated values should still add to the original sum; a missing value in one component that shows good dependence or association with some other nonmissing components should be replaced by a mean estimate conditional on those other components.

Replacements can be either fixed or random. Fixed replacements are easy to implement and manage. Unfortunately, they implicitly reduce the natural variability of the data set. On the contrary, random replacements are a bit more complex and require an idea of how and why the datum was lost, that is, a model of randomness to use.

In summary, for a small number of missing values, a replacement by a fixed, reasonable value can be a sensible solution, which allows to apply the whole set of tools and methods of this book. If the number of missing values is large, random replacements, or any of the other strategies, should be considered.

Replacement strategies include additive replacement (Aitchison, 1986); multiplicative replacement (Martín-Fernández et al., 2003; Martín-Fernández et al., 2011), eventually complemented with randomness (Martín-Fernández et al., 2003); conditional mean replacement for rounded zeros and for missing values (Palarea-Albaladejo et al., 2007; Palarea-Albaladejo and Martín-Fernández, 2008; Hron et al., 2010; Boogaart and Tolosana-Delgado, 2013); and Box–Cox-based replacement, minimizing subcompositional incoherence (Greenacre, 2010, 2011).

Subcompositional analysis

It might be that a value cannot be replaced by anything sensible, simply because the real value *must be* zero (a structural zero) or no information at all is available on it. Or a variable might be so often missing, that it is unreliable to extract from the nonmissing part any information to use in the replacement. In these cases, it might be more reasonable to consider that variable as a factor, converting it into a 0 − 1 variable (missing/present), and use the techniques presented in Sections 7.3 or 8.3 to check for differences on the subcomposition analysis without the affected component (Bacon-Shone, 2003). An illustration of this methodology is given in Section 5.9.

A related technique, the exhaustive use of all pairwise logratios, is based on the variation matrix concept (see Definition 5.2). It consists of computing all possible pairwise logratios separately and then looking for a compatible representation of results. This strategy has been applied to geostatistics (Tolosana-Delgado et al., 2009) and regression with compositional response (Tolosana-Delgado and Eynatten, 2009). Interested readers might find more information in the last paper or in Boogaart and Tolosana-Delgado (2013).

Latent compositions

This is a wide class of methods that can be applied to many data sets, not necessarily with zeros. It assumes the existence of a nonobserved (or *latent*) random composition without missing values and considers the observed data to be a known function of this latent composition. Then, one analyzes the latent composition through the information that can be derived from the observations, which might not be compositional themselves (i.e., not fulfilling the principles mentioned earlier). Such a strategy can be useful for all kinds of irregular data, compositions from counts (Daunis-i-Estadella et al., 2008), mixtures (Tolosana-Delgado et al., 2011), and so on. But, on the down side, this strategy requires strong models linking the observations with the latent composition. Moreover, solutions are mostly only available through likelihood-based approaches (maximum likelihood, Bayesian analysis, see Section 7.5), which are not adequate for exploratory analysis.

2.4   Exercises

Exercise 1. If data are measured in ppm, what is the value of the constant κ in Definition 2.4?

Table 2.1   Simulated data set (three parts, 20 samples)

	1	2	3	4	5	6	7	8	9	10
x_1	79.07	31.74	18.61	49.51	29.22	21.99	11.74	24.47	5.14	15.54
x_2	12.83	56.69	72.05	15.11	52.36	59.91	65.04	52.53	38.39	57.34
x_3	8.10	11.57	9.34	35.38	18.42	18.10	23.22	23.00	56.47	27.11

	11	12	13	14	15	16	17	18	19	20
x_1	57.17	52.25	77.40	10.54	46.14	16.29	32.27	40.73	49.29	61.49
x_2	3.81	23.73	9.13	20.34	15.97	69.18	36.20	47.41	42.74	7.63
x_3	39.02	24.02	13.47	69.12	37.89	14.53	31.53	11.86	7.97	30.88

Exercise 2. Plot a ternary diagram using different values for the constant sum κ.

Exercise 3. Verify that data in Table 2.1 satisfy the conditions for being compositional. Plot them in a ternary diagram.

Exercise 4. Compute the Euclidean distance between the first two vectors of Table 2.1. Imagine originally a fourth variable x_4 was measured, constant for all samples and equal to 5%. Take the first two vectors, close them to sum up to 95%, add the fourth variable to them (so that they sum up to 100%), and compute the Euclidean distance between the closed vectors. If the Euclidean distance is subcompositionally dominant, the distance between the four-part compositions must be greater than or equal to the distance between the three-part compositions.

3

The Aitchison geometry

3.1 General comments

In real space, we are used to add vectors, to multiply them by a constant or scalar value, to look for properties such as orthogonality, or to compute the distance between two points. All this, and much more, is possible because the real space is a linear vector space with a metric structure. We are familiar with its geometric structure, the Euclidean geometry, and we are used to represent our observations within this geometry. But this geometry is not a proper geometry for compositional data.

To illustrate this assertion, consider the compositions $[5, 65, 30]$, $[10, 60, 30]$, $[50, 20, 30]$, and $[55, 15, 30]$. Intuitively, we would say that the difference between $[5, 65, 30]$ and $[10, 60, 30]$ is not the same as the difference between $[50, 20, 30]$ and $[55, 15, 30]$. The Euclidean distance between them is certainly the same, as there is a difference of 5 units both between the first and the second respective components. But in the first case, the proportion in the first component is doubled, while in the second case, the relative increase is about 10%. This relative difference seems more adequate to describe compositional variability.

This is not the only reason for discarding the usual Euclidean geometry as a proper tool for analyzing compositional data. Problems might appear in many situations, such as those where results end up outside the sample space, for example, when translating compositional vectors, or computing joint confidence

Modeling and Analysis of Compositional Data, First Edition.
Vera Pawlowsky-Glahn, Juan José Egozcue and Raimon Tolosana-Delgado.

regions for random compositions under assumptions of normality, or using hexagonal fields of variation (Stevens et al., 1956). This last case is paradigmatic, as such hexagons are often naively cut when they lay partly outside the ternary diagram, and this without regard to any probability adjustment. These types of problems are not just theoretical: they are practical and interpretative. Many examples can be found in the published literature (e.g., Ingersoll, 1978; Ingersoll and Suczek, 1979).

What is needed is a sensible geometry to work with compositional data. In the simplex, things appear not as simple as they (apparently) are in real space, but ways can be found to work in the simplex that are completely analogous to working in the real space. In fact, it is possible to define two operations that give the simplex a vector space structure. The first one is perturbation, which is analogous to addition in real space; the second one is powering, which is analogous to multiplication by a scalar in real space. Both require in their definition the closure operation; recall that, as stated in Definitions 2.2 and 2.3, closure is nothing else but the projection of a vector with positive components onto the simplex. Moreover, it is possible to obtain a Euclidean vector space structure on the simplex, just adding an inner product, a norm, and a distance to the previous definitions. With the inner product, compositions can be projected onto particular directions and one can check for orthogonality and also determine angles between compositional vectors; with the norm, the *length* of a composition can be computed; the possibilities of a distance should be clear. With all these together, one can operate in the simplex in the same way as one operates in real space.

3.2   Vector space structure

The basic operations required for a vector space structure of the simplex follow. They use the closure operation given in Definition 2.3.

Definition 3.1 *(Perturbation).*
Perturbation of $\mathbf{x} \in S^D$ *by* $\mathbf{y} \in S^D$,

$$\mathbf{x} \oplus \mathbf{y} = C[x_1 y_1, x_2 y_2, \dots, x_D y_D] \in S^D.$$

Definition 3.2 *(Powering).*
Power transformation or powering of $\mathbf{x} \in S^D$ *by a constant* $\alpha \in \mathbb{R}$,

$$\alpha \odot \mathbf{x} = C[x_1^\alpha, x_2^\alpha, \dots, x_D^\alpha] \in S^D.$$

For illustration of the effect of perturbation and powering, see Figure 3.1.

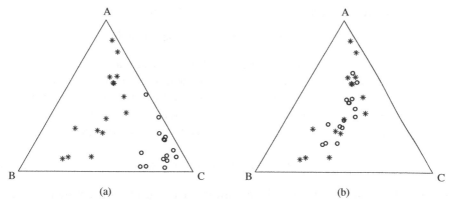

Figure 3.1   (a) Perturbation of initial compositions (∘) by **p** *= [0.1, 0.1, 0.8] resulting in compositions (*). (b) Powering of compositions (*) by* α *= 0.5 resulting in compositions (∘).*

The simplex, $(S^D, \oplus, \odot)$, with perturbation and powering, is a vector space. This means the following properties hold, making them analogous to translation and scalar multiplication in real space.

Property 3.3 $(S^D, \oplus)$ is a commutative group structure; that is, for $\mathbf{x}, \mathbf{y}, \mathbf{z} \in S^D$, it holds

1. commutative property: $\mathbf{x} \oplus \mathbf{y} = \mathbf{y} \oplus \mathbf{x}$;

2. associative property: $(\mathbf{x} \oplus \mathbf{y}) \oplus \mathbf{z} = \mathbf{x} \oplus (\mathbf{y} \oplus \mathbf{z})$;

3. neutral element:

$$\mathbf{n} = C[1, 1, \ldots, 1] = \left[\frac{1}{D}, \frac{1}{D}, \ldots, \frac{1}{D}\right];$$

$\mathbf{n}$ is the barycenter of the simplex and is unique;

4. inverse of $\mathbf{x}$: $\mathbf{x}^{-1} = C\left[x_1^{-1}, x_2^{-1}, \ldots, x_D^{-1}\right]$; thus, $\mathbf{x} \oplus \mathbf{x}^{-1} = \mathbf{n}$.

By analogy with standard operations in real space, we will write $\mathbf{x} \oplus \mathbf{y}^{-1} = \mathbf{x} \ominus \mathbf{y}$ for the perturbation difference.

Property 3.4 Powering satisfies the properties of an external product. For $\mathbf{x}, \mathbf{y} \in S^D$, $\alpha, \beta \in \mathbb{R}$, it holds

1. associative property: $\alpha \odot (\beta \odot \mathbf{x}) = (\alpha \cdot \beta) \odot \mathbf{x}$;

2. distributive property 1: $\alpha \odot (\mathbf{x} \oplus \mathbf{y}) = (\alpha \odot \mathbf{x}) \oplus (\alpha \odot \mathbf{y})$;

3. distributive property 2: $(\alpha + \beta) \odot \mathbf{x} = (\alpha \odot \mathbf{x}) \oplus (\beta \odot \mathbf{x})$;

4. neutral element: $1 \odot \mathbf{x} = \mathbf{x}$; the neutral element is unique.

Note that the closure operation cancels out any constant and, thus, the closure constant itself is not important from a mathematical point of view. This fact allows us to omit the closure in intermediate steps of any computation without problem. It has also significant implications for practical reasons, as shall be seen during simplicial principal component analysis. We can express this property for $\mathbf{z} \in \mathbb{R}_+^D$ and $\mathbf{x} \in S^D$ as

$$\mathbf{x} \oplus (\alpha \odot \mathbf{z}) = \mathbf{x} \oplus (\alpha \odot C(\mathbf{z})). \tag{3.1}$$

Nevertheless, one should be always aware that the closure constant is very important for the interpretation of the units of the problem at hand. Therefore, controlling for the right units should be the last step in any analysis.

3.3   Inner product, norm and distance

To obtain a Euclidean vector space structure, we take the following inner product, with associated norm and distance (the subindex a stands for Aitchison).

Definition 3.5 *(Aitchison inner product).*
Inner product of $\mathbf{x}, \mathbf{y} \in S^D$,

$$\langle \mathbf{x}, \mathbf{y} \rangle_a = \frac{1}{2D} \sum_{i=1}^{D} \sum_{j=1}^{D} \ln \frac{x_i}{x_j} \ln \frac{y_i}{y_j}.$$

Definition 3.6 *(Aitchison norm).*
Norm of $\mathbf{x} \in S^D$,

$$\|\mathbf{x}\|_a = \sqrt{\frac{1}{2D} \sum_{i=1}^{D} \sum_{j=1}^{D} \left(\ln \frac{x_i}{x_j} \right)^2}.$$

Definition 3.7 *(Aitchison distance).*
Distance between $\mathbf{x}$ *and* $\mathbf{y} \in S^D$,

$$d_a(\mathbf{x}, \mathbf{y}) = \|\mathbf{x} \ominus \mathbf{y}\|_a = \sqrt{\frac{1}{2D} \sum_{i=1}^{D} \sum_{j=1}^{D} \left(\ln \frac{x_i}{x_j} - \ln \frac{y_i}{y_j} \right)^2}.$$

In practice, alternative, but equivalent, expressions of the inner product, norm, and distance are useful. Three possible alternatives for the inner product are as follows:

$$\langle \mathbf{x}, \mathbf{y} \rangle_a = \frac{1}{D} \sum_{i=1}^{D-1} \sum_{j=i+1}^{D} \ln \frac{x_i}{x_j} \ln \frac{y_i}{y_j} \tag{3.2}$$

$$= \sum_{i=1}^{D} \ln x_i \ln y_i - \frac{1}{D} \left(\sum_{j=1}^{D} \ln x_j \right) \left(\sum_{k=1}^{D} \ln y_k \right) \tag{3.3}$$

$$= \sum_{i=1}^{D} \ln \frac{x_i}{g_m(\mathbf{x})} \cdot \ln \frac{y_i}{g_m(\mathbf{y})}, \tag{3.4}$$

where $g_m(\cdot)$ denotes the geometric mean of the arguments. Expression (3.4) corresponds to an ordinary inner product of two real vectors. These vectors are called centered logratio (clr) of $\mathbf{x}, \mathbf{y}$, as defined in Chapter 4. Note that simple logratios, $\ln(x_i/x_j)$, are null whenever $i = j$; therefore, in Equation (3.2),

$$\sum_{i=1}^{D-1} \sum_{j=i+1}^{D} \ln \frac{x_i}{x_j} \ln \frac{y_i}{y_j} = \frac{1}{2} \sum_{i=1}^{D} \sum_{j=1}^{D} \ln \frac{x_i}{x_j} \ln \frac{y_i}{y_j}.$$

To refer to the properties of $(S^D, \oplus, \odot)$ as an Euclidean linear vector space, we shall talk globally about the *Aitchison geometry on the simplex* and, in particular, about the Aitchison distance, norm, and inner product. In mathematical textbooks, such a linear vector space is called either real Euclidean space or finite dimensional real Hilbert space.

The algebraic-geometric structure of S^D satisfies standard properties, such as compatibility of the distance with perturbation and powering, that is,

$$d_a(\mathbf{p} \oplus \mathbf{x}, \mathbf{p} \oplus \mathbf{y}) = d_a(\mathbf{x}, \mathbf{y}), \quad d_a(\alpha \odot \mathbf{x}, \alpha \odot \mathbf{y}) = |\alpha| d_a(\mathbf{x}, \mathbf{y}),$$

for any $\mathbf{x}, \mathbf{y}, \mathbf{p} \in S^D$, and $\alpha \in \mathbb{R}$. Other typical properties of Euclidean spaces are valid for S^D. Some of them are as follows:

1. Cauchy–Schwartz inequality:

$$|\langle \mathbf{x}, \mathbf{y} \rangle_a| \le \|\mathbf{x}\|_a \cdot \|\mathbf{y}\|_a;$$

2. Pythagoras theorem: If $\mathbf{x}, \mathbf{y}$ are orthogonal, that is, $\langle \mathbf{x}, \mathbf{y} \rangle_a = 0$, then

$$\|\mathbf{x} \ominus \mathbf{y}\|_a^2 = \|\mathbf{x}\|_a^2 + \|\mathbf{y}\|_a^2;$$

3. Triangular inequality:

$$d_a(\mathbf{x}, \mathbf{y}) \leq d_a(\mathbf{x}, \mathbf{z}) + d_a(\mathbf{y}, \mathbf{z}).$$

For a discussion of these and other properties, see Billheimer et al. (2001) or Pawlowsky-Glahn and Egozcue (2001). For a comparison with other measures of difference obtained as restrictions of distances in $\mathbb{R}^D$ to S^D, see Martín-Fernández et al. (1998, 1999), Aitchison et al. (2000), or Martín-Fernández (2001).

The Aitchison distance is subcompositionally coherent, as perturbation (Definition 3.1), powering (Definition 3.2), and inner product (Definition 3.5) induce the same linear vector space structure in the subspace corresponding to a subcomposition. Moreover, the distance is subcompositionally dominant, as illustrated by Exercise 11.

3.4   Geometric figures

Within this geometric framework, we can define lines in S^D, which we call *compositional lines*, as $\mathbf{y} = \mathbf{x}_0 \oplus (\alpha \odot \mathbf{x})$, with $\mathbf{x}_0$ the starting point, $\mathbf{x}$ the leading vector, and α a real parameter. Note that $\mathbf{y}$, $\mathbf{x}_0$, and $\mathbf{x}$ are elements of S^D, while the coefficient α varies in $\mathbb{R}$. To illustrate what we understand by *compositional lines*, Figure 3.2 shows two families of parallel lines in a ternary diagram, forming a square, orthogonal, grid of side equal to one Aitchison distance unit. Recall that parallel lines have the same leading vector, but different starting points, for instance,

$$\mathbf{y}_1 = \mathbf{x}_1 \oplus (\alpha \odot \mathbf{x}) \quad \text{and} \quad \mathbf{y}_2 = \mathbf{x}_2 \oplus (\alpha \odot \mathbf{x}),$$

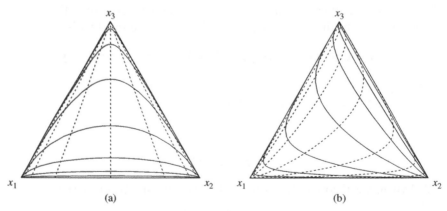

(a)   (b)

Figure 3.2   Orthogonal grids of compositional lines in S^3, equally spaced 1 unit in the Aitchison distance (Definition 3.7). The grid in (a) is rotated 45° clockwise with respect to the grid in (b). Each family has seven lines.

while orthogonal lines are those for which the inner product of the leading vectors is zero, that is, for

$$\mathbf{y}_1 = \mathbf{x}_0 \oplus (\alpha_1 \odot \mathbf{x}_1) \quad \text{and} \quad \mathbf{y}_2 = \mathbf{x}_0 \oplus (\alpha_2 \odot \mathbf{x}_2),$$

with $\mathbf{x}_0$ their intersection point and $\mathbf{x}_1$, $\mathbf{x}_2$ the corresponding leading vectors. In this case, it holds $\langle \mathbf{x}_1, \mathbf{x}_2 \rangle_a = 0$. Thus, *orthogonal* means here that the inner product given in Definition 3.5 of the leading vectors of two lines, one of each family, is zero, and one Aitchison distance unit is measured by the distance given in Definition 3.7.

Once a well-defined geometry is available, it is straightforward to define any geometric figure, for instance, parallel lines and rhomboids (Figure 3.3) or circles and ellipses (Figure 3.4). Figure 3.3a shows two pairs of parallel lines forming a rhomboid, that is, they are not orthogonal. Note that this figure (and the next ones) uses a reference axis (dotted lines) equivalent to that of Figure 3.2a, although the whole figure (including labels) has been actually rotated 120° clockwise. A rhomboid (Figure 3.3b) can also be constructed by perturbing a line segment (thick line), that is, the set of points between $\mathbf{a}$ and $\mathbf{b}$ along the leading composition $\mathbf{v} = \mathbf{b} \ominus \mathbf{a}$. The perturbing vector is shown in that figure as a dashed segment from the barycenter to $\mathbf{p}$. The final perturbed segment is thus the line between $\mathbf{a} \oplus \mathbf{p}$ and $\mathbf{b} \oplus \mathbf{p}$ along $\mathbf{v}$.

A compositional circle is the set of compositions placed at a constant Aitchison distance r from a central composition $\mathbf{x}_0$, that is, the set of compositions $\mathbf{x}$ satisfying $d_a(\mathbf{x}, \mathbf{x}_0) = r$. Figure 3.4 shows a set of three concentric circles around $\mathbf{x}_0 = \mathbf{n}$, with radii $r = 0.2$, $r = 0.4$, and $r = 0.6$. The existence of a Euclidean

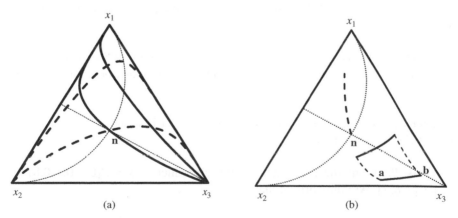

Figure 3.3   (a) Parallel lines in S^3, and (b) construction of a rhomboid by perturbation of a segment.

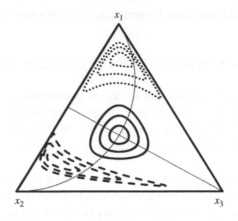

Figure 3.4   Circles and ellipses in S^3.

space structure allows to construct other, more complex geometric figures, such as the ellipses shown in this same figure. However, their parameterization requires advanced geometric concepts that are easier to handle in coordinate representation, introduced in Chapter 4.

3.5   Exercises

Exercise 5.   Consider the two vectors $[0.7, 0.5, 0.8]$ and $[0.25, 0.75, 0.5]$. Perturb one vector by the other with and without previous closure. Is there any difference?

Exercise 6.   Perturb each sample of the data set given in Table 2.1 with $s = C[0.1, 0.1, 0.8]$ and plot the initial and the resulting perturbed data set. What do you observe?

Exercise 7.   Apply powering with α ranging from -8 to $+8$ in steps of 1 unit to $t = C[0.7, 0.5, 0.8]$ and plot the resulting set of compositions. Join them by a line. What do you observe?

Exercise 8.   Perturb the compositions obtained in Exercise 7 by $r = C[0.25, 0.75, 0.5]$. What is the result?

Exercise 9.   Compute the Aitchison inner product of $x = C[0.7, 0.4, 0.8]$ and $y = C[2, 8, 1]$. Are they orthogonal?

Exercise 10. Compute the Aitchison norm of $\mathbf{x}_1 = C[0.7, 0.4, 0.8]$ and call it a. Compute $\alpha \odot \mathbf{x}_1$ with $\alpha = 1/a$. Compute the Aitchison norm of the resulting composition. How do you interpret the result?

Exercise 11. Redo Exercise 4, but using the Aitchison distance given in Definition 3.7. Is it subcompositionally dominant?

Exercise 12. In a two-part composition $\mathbf{x} = [x_1, x_2]$, simplify the formula for the Aitchison distance, taking $x_2 = 1 - x_1$ (using $\kappa = 1$). Use it to plot seven equally spaced points on the segment $(0, 1) = S^2$, from $x_1 = 0.014$ to $x_1 = 0.986$.

Exercise 13. In a rock sample, several radioactive isotopes have been measured, obtaining $[^{238}U, {}^{232}Th, {}^{40}K] = [150, 30, 110]$ ppm. Which will be the composition after $\Delta t = 10^9$ years? And after another Δt years? Which was the composition Δt years ago? And Δt years before that? Close these five compositions and represent them in a ternary diagram. What do you see? Could you write the evolution as an equation? (Half-life disintegration periods: $[^{238}U, {}^{232}Th, {}^{40}K] = [4.468; 14.05; 1.277] \cdot 10^9$ years).

4

Coordinate representation

4.1 Introduction

To introduce transformations based on ratios, Aitchison (1986) used the fact that, for compositional data, size is irrelevant – as interest lies in relative proportions of the components measured. The essential transformations introduced by him are the additive logratio transformation (alr) and the centered logratio transformation (clr). He applied classical statistical analysis to the transformed observations, using the alr transformation for modeling, and the clr transformation for those techniques based on a metric. Although not explicit, the underlying reason was that the alr transformation does not preserve distances, whereas the clr transformation preserves distances but leads to a singular covariance matrix. In mathematical terms, we say that the alr transformation is an isomorphism between S^D and $\mathbb{R}^{D-1}$, but not an isometry. The clr transformation is an isometry, and thus also an isomorphism, but between S^D and a subspace of $\mathbb{R}^D$, leading to degenerate distributions. Thus, Aitchison's approach opened up a rigorous strategy, but care had to be applied when using either of the transformations.

Using the Euclidean vector space structure of the simplex, it is possible to give an algebraic-geometric foundation to his approach, and it is possible to go even a step further. Within this framework, a transformation of coefficients is equivalent to express observations in a different coordinate system. At school we are taught to work in an orthogonal system, known as a Cartesian coordinate system;

Modeling and Analysis of Compositional Data, First Edition.
Vera Pawlowsky-Glahn, Juan José Egozcue and Raimon Tolosana-Delgado.
© 2015 John Wiley & Sons, Ltd. Published 2015 by John Wiley & Sons, Ltd.

we learn how to change coordinates within this system and how to rotate axis. But neither the clr nor the alr transformations can be directly associated with an orthogonal coordinate system in the simplex, a fact that leads to define a new transformation, called ilr (for *isometric logratio*) transformation, which is an isometry between S^D and $\mathbb{R}^{D-1}$, thus avoiding the drawbacks of both the alr and the clr (Egozcue et al., 2003). The term ilr stands actually for the association of coordinates with compositions in an orthonormal system in general, and this is the framework presented here, together with a particular kind of coordinates, named balances, because of their usefulness for modeling and interpretation.

4.2   Compositional observations in real space

Compositions in S^D are usually expressed in terms of the canonical basis of $\mathbb{R}^D$, $\{\mathbf{e}_1, \mathbf{e}_2, \dots, \mathbf{e}_D\}$. In fact, any vector $\mathbf{x} \in \mathbb{R}^D$ can be written as

$$\mathbf{x} = x_1[1, 0, \dots, 0] + x_2[0, 1, \dots, 0] + \dots + x_D[0, 0, \dots, 1] = \sum_{i=1}^{D} x_i \cdot \mathbf{e}_i, \quad (4.1)$$

and this is the way we are used to interpret it. The problem is that the set of vectors $\{\mathbf{e}_1, \mathbf{e}_2, \dots, \mathbf{e}_D\}$ is neither a generating system nor a basis with respect to the vector space structure of S^D defined in Chapter 3. In fact, not every combination of coefficients gives an element of S^D (negative and zero values are not allowed), and the $\mathbf{e}_i$ themselves do not belong to the simplex as defined in Equation (2.1). Nevertheless, in many cases, it is interesting to express results in terms of compositions as given by Equation (4.1), so that interpretations are feasible in usual units. Therefore, one of our purposes is to find a way to state statistically rigorous results in this coordinate system.

4.3   Generating systems

A first step for defining an appropriate orthonormal basis consists of finding a generating system that can be used to build the basis. A natural way to obtain such a generating system is to take exponentials in the canonical basis of Section 4.2, resulting in $\{\mathbf{w}_1, \mathbf{w}_2, \dots, \mathbf{w}_D\}$, with

$$\mathbf{w}_i = C(\exp(\mathbf{e}_i)) = C[1, 1, \dots, e, \dots, 1], \quad i = 1, 2, \dots, D. \quad (4.2)$$

Here the number e is placed in the ith column of $\mathbf{w}_i$ and $\exp(\cdot)$ is assumed to operate componentwise. In fact, taking into account Equation (3.1) and the usual rules of precedence for operations in a vector space, that is, first the external operation, $\odot$, and afterwards the internal operation, $\oplus$, any vector $\mathbf{x} \in S^D$ can be written as

$$\mathbf{x} = \bigoplus_{i=1}^{D} \ln x_i \odot \mathbf{w}_i$$
$$= \ln x_1 \odot [e, 1, \dots, 1] \oplus \ln x_2 \odot [1, e, \dots, 1] \oplus \dots \oplus \ln x_D \odot [1, 1, \dots, e].$$

In this expression, the closure of each individual vector has been removed, as each perturbation and each powering includes it (see Exercise 5). Coefficients with respect to a generating system are not unique. Thus, the following equivalent expression can be used as well,

$$\mathbf{x} = \bigoplus_{i=1}^{D} \ln \frac{x_i}{g_m(\mathbf{x})} \odot \mathbf{w}_i$$
$$= \ln \frac{x_1}{g_m(\mathbf{x})} \odot [e, 1, \dots, 1] \oplus \dots \oplus \ln \frac{x_D}{g_m(\mathbf{x})} \odot [1, 1, \dots, e],$$

where

$$g_m(\mathbf{x}) = \left(\prod_{i=1}^{D} x_i \right)^{1/D} = \exp\left(\frac{1}{D} \sum_{i=1}^{D} \ln x_i \right)$$

is the componentwise geometric mean of the composition. In this second expression, the coefficients of the centered logratio transformation defined by Aitchison (1986) can be recognized. Note that the denominator could be replaced *by any constant*. This nonuniqueness is consistent with the concept of compositions as equivalence classes as given in Definition 2.2 (Barceló-Vidal et al., 2001).

In what follows, clr will denote the transformation that gives the expression of a composition in centered logratio coefficients

$$\mathrm{clr}(\mathbf{x}) = \left[\ln \frac{x_1}{g_m(\mathbf{x})}, \ln \frac{x_2}{g_m(\mathbf{x})}, \dots, \ln \frac{x_D}{g_m(\mathbf{x})} \right] = \xi. \qquad (4.3)$$

The inverse transformation, which gives the coefficients in the canonical basis of real space, is then

$$\mathrm{clr}^{-1}(\xi) = C[\exp(\xi_1), \exp(\xi_2), \dots, \exp(\xi_D)] = \mathbf{x}. \qquad (4.4)$$

The centered logratio transformation is symmetric in the components, but the price is a new constraint on the transformed sample: the sum of the components is zero. This means that the transformed sample lies on a plane, which goes through the origin of $\mathbb{R}^D$, and is orthogonal to the vector of unities, $[1, 1, \dots, 1]$. But, more importantly, it means also that for random compositions (see Chapter 6), the covariance matrix of clr($\mathbf{x}$) = ξ is singular, that is, the determinant is zero. Furthermore, clr coefficients are not subcompositionally coherent, because the geometric mean of the parts of a subcomposition $g_m(\mathbf{x}_s)$ is not necessarily equal to that of the full composition, and thus the clr coefficients are in general not the same. The following is a formal definition of the clr coefficients.

Definition 4.1 (Centered logratio coefficients).
For a composition $\mathbf{x} \in S^D$, *the* clr *coefficients are the components of the unique vector* $\xi = [\xi_1, \xi_2, \ldots, \xi_D] =$ clr($\mathbf{x}$), *satisfying the two conditions*

$$\mathbf{x} = \text{clr}^{-1}(\xi) = C(\exp(\xi)) \quad and \quad \sum_{i=1}^{D} \xi_i = 0.$$

The ith clr *coefficient is*

$$\xi_i = \ln \frac{x_i}{g_m(\mathbf{x})},$$

with $g_m(\mathbf{x})$ *the geometric mean of the components of* $\mathbf{x}$.

The clr coefficients are not coordinates with respect to a basis of the simplex, but they have very important properties. Among them, the translation of operations and metrics from the simplex into real space deserves special attention. Denote ordinary (Euclidean) distance, norm, and inner product in $\mathbb{R}^{D-1}$ by $d(\cdot, \cdot)$, $\| \cdot \|$, and $\langle \cdot, \cdot \rangle$, respectively. The following properties hold.

Property 4.2 Consider $\mathbf{x}_k \in S^D$ and real constants α, β; then for all $k = 1, 2$, it holds that

(a) clr($\alpha \odot \mathbf{x}_1 \oplus \beta \odot \mathbf{x}_2$) = $\alpha \cdot$ clr($\mathbf{x}_1$) + $\beta \cdot$ clr($\mathbf{x}_2$);

(b) $\langle \mathbf{x}_1, \mathbf{x}_2 \rangle_a = \langle$clr($\mathbf{x}_1$), clr($\mathbf{x}_2$)$\rangle$;

(c) $\|\mathbf{x}_1\|_a = \|$clr($\mathbf{x}_1$)$\|$, $d_a(\mathbf{x}_1, \mathbf{x}_2) = d($clr($\mathbf{x}_1$), clr($\mathbf{x}_2$)).

Property 4.2 points out that clr : $S^D \rightarrow U \subset \mathbb{R}^D$ is a mapping of the simplex on a $(D-1)$-dimensional subspace U, where U contains all real vectors which components add to zero. Property (a) tells us that the clr is an isomorphism of vector spaces: operations in the simplex are translated into operations in a real vector space. Properties (b) and (c) characterize clr as an isometry: distances, orthogonal projections, norms, and all metric concepts in the simplex are maintained after a clr-transformation.

A word of caution is necessary here: properties stated above hold for the whole clr-vector, but not pairwise for its components, as they are not orthonormal coordinates.

4.4   Orthonormal coordinates

Omitting a composition of the generating system given in Equation (4.2), a basis is obtained. For example, omitting $\mathbf{w}_D$ results in $\{\mathbf{w}_1, \mathbf{w}_2, \dots, \mathbf{w}_{D-1}\}$. This basis is not orthonormal, as can be shown by computing the inner product of any two of its vectors. But a new basis, orthonormal with respect to the inner product, can be readily obtained using the Gram–Schmidt procedure (Egozcue et al., 2003). The basis thus obtained will be just one out of the infinitely many orthonormal basis that can be defined in any Euclidean space. Therefore, it is convenient to study their general characteristics. To do so, we need the following definition.

Definition 4.3 (Contrast matrix).
Let $\{\mathbf{e}_1, \mathbf{e}_2, \dots, \mathbf{e}_{D-1}\}$ be an orthonormal basis of the simplex S^D and consider the $(D-1, D)$-matrix $\boldsymbol{\Psi} = [\psi_{ij}]$ in which rows are the $\boldsymbol{\psi}_i = \mathrm{clr}(\mathbf{e}_i)$, $i = 1, 2, \dots,$ $D-1$. This matrix is called contrast matrix associated with the orthonormal basis $\{\mathbf{e}_1, \mathbf{e}_2, \dots, \mathbf{e}_{D-1}\}$. Each row is called a (log)contrast.

An orthonormal basis satisfies that

$$\langle \mathbf{e}_i, \mathbf{e}_j \rangle_a = \delta_{ij},$$

where δ_{ij} stands for the Kronecker-delta, which is null for $i \neq j$, and one whenever $i = j$. This can be expressed using Property 4.2(b),

$$\langle \mathbf{e}_i, \mathbf{e}_j \rangle_a = \langle \mathrm{clr}(\mathbf{e}_i), \mathrm{clr}(\mathbf{e}_j) \rangle = \delta_{ij}.$$

It implies that the $(D-1, D)$-matrix $\boldsymbol{\Psi}$ satisfies

$$\boldsymbol{\Psi}\boldsymbol{\Psi}^\top = \mathbf{I}_{D-1}, \quad \boldsymbol{\Psi}^\top\boldsymbol{\Psi} = \mathbf{I}_D - \left(\frac{1}{D}\right) \mathbf{1}_D^\top \mathbf{1}_D, \tag{4.5}$$

where $\mathbf{I}_D$, $\mathbf{I}_{D-1}$ are the D and $(D-1)$-dimensional identity matrices; $^\top$ denotes transposition; and $\mathbf{1}_D$ is a D row vector of ones. Note that $\boldsymbol{\Psi}$ is a matrix of rank $D-1$. The compositions of the basis are recovered from $\boldsymbol{\Psi}$ using clr^{-1} (Equation 4.4) in each row of the matrix. Recall that the rows of $\boldsymbol{\Psi}$ are clr coefficients and therefore, as stated in Definition 4.1, they add up to 0.

Once an orthonormal basis has been chosen, a composition $\mathbf{x} \in S^D$ is expressed as

$$\mathbf{x} = \bigoplus_{i=1}^{D-1} x_i^* \odot \mathbf{e}_i, \quad x_i^* = \langle \mathbf{x}, \mathbf{e}_i \rangle_a, \tag{4.6}$$

where $\mathbf{x}^* = \begin{bmatrix} x_1^*, x_2^*, \ldots, x_{D-1}^* \end{bmatrix}$ is the vector of coordinates of $\mathbf{x}$ with respect to the selected basis.

Definition 4.4 (Isometric logratio transformation and coordinates).
Let $\{\mathbf{e}_1, \mathbf{e}_2, \ldots, \mathbf{e}_{D-1}\}$ *be an orthonormal basis of the simplex* S^D. *The isometric logratio transformation, ilr for short, of the composition* $\mathbf{x}$ *is the function* ilr : $S^D \to \mathbb{R}^{D-1}$, *which assigns the coordinates* $\mathbf{x}^*$, *with respect to the given basis, to the composition* $\mathbf{x}$. *The vector* $\mathbf{x}^*$ *contains the* $D - 1$ *ilr-coordinates of* $\mathbf{x}$. *The inverse of the* ilr-*transformation is denoted as* ilr^{-1}.

After this general definition, the properties of the ilr-coordinates explain both their potential application as well as their computation. The function ilr : $S^D \to \mathbb{R}^{D-1}$ is an isometry of vector spaces. In what follows, the asterisk (*) is used to denote coordinates in an orthonormal basis.

Property 4.5 Consider $\mathbf{x}_k \in S^D$, $k = 1, 2$, and real constants α, β; then

(a) $\text{ilr}(\alpha \odot \mathbf{x}_1 \oplus \beta \odot \mathbf{x}_2) = \alpha \cdot \text{ilr}(\mathbf{x}_1) + \beta \cdot \text{ilr}(\mathbf{x}_2) = \alpha \cdot \mathbf{x}_1^* + \beta \cdot \mathbf{x}_2^*$;

(b) $\langle \mathbf{x}_1, \mathbf{x}_2 \rangle_a = \langle \text{ilr}(\mathbf{x}_1), \text{ilr}(\mathbf{x}_2) \rangle = \langle \mathbf{x}_1^*, \mathbf{x}_2^* \rangle$;

(c) $\|\mathbf{x}_1\|_a = \|\text{ilr}(\mathbf{x}_1)\| = \|\mathbf{x}_1^*\|$;

(d) $d_a(\mathbf{x}_1, \mathbf{x}_2) = d(\text{ilr}(\mathbf{x}_1), \text{ilr}(\mathbf{x}_2)) = d(\mathbf{x}_1^*, \mathbf{x}_2^*)$.

The main difference between Property 4.2 for clr and Property 4.5 for ilr is that the former refers to vectors of coefficients in $\mathbb{R}^D$, while the latter deals with vectors of coordinates in $\mathbb{R}^{D-1}$, thus matching the dimension of S^D.

Taking into account Properties 4.2 and 4.5, and using the contrast matrix $\boldsymbol{\Psi}$ of the basis (Definition 4.3), the coordinates of a composition $\mathbf{x}$ can be expressed in a compact way. As written in Equation (4.6), a coordinate can be obtained as the result of an Aitchison inner product and can thus be expressed as an ordinary inner product of the clr coefficients. Grouping all coordinates in a vector,

$$\mathbf{x}^* = \text{ilr}(\mathbf{x}) = \text{clr}(\mathbf{x}) \cdot \boldsymbol{\Psi}^\mathsf{T} = \ln(\mathbf{x}) \cdot \boldsymbol{\Psi}^\mathsf{T}, \qquad (4.7)$$

the ilr transformation is computed as a simple matrix product. Inversion of ilr, that is, recovering the composition from its coordinates, corresponds to Equation (4.6). In fact, taking clr coefficients in both sides of Equation (4.6) and taking into account Property 4.2, it holds

$$\text{clr}(\mathbf{x}) = \mathbf{x}^* \boldsymbol{\Psi}, \qquad \mathbf{x} = C(\exp(\mathbf{x}^* \boldsymbol{\Psi})) = \text{ilr}^{-1}(\mathbf{x}^*). \qquad (4.8)$$

A simple algorithm to recover $\mathbf{x}$ from its coordinates $\mathbf{x}^*$ consists of the following steps:

1. construct the contrast matrix of the basis, $\boldsymbol{\Psi}$;

2. compute the matrix product $\mathbf{x}^*\boldsymbol{\Psi}$;

3. apply clr^{-1} to obtain $\mathbf{x}$.

4.5   Balances

There are several ways to define orthonormal bases in the simplex. The main criterion for the selection of an orthonormal basis is that it should enhance interpretability of the representation in coordinates. For instance, when performing principal component analysis, an orthogonal basis is selected so that the first coordinate (principal component) represents the direction of maximum variability. Particular cases deserving our attention are those bases linked to a sequential binary partition (SBP) of the compositional vector (Egozcue and Pawlowsky-Glahn 2005; Pawlowsky-Glahn and Egozcue 2011). The main interest of such bases is that they are easily interpreted in terms of grouped parts of the composition. The Cartesian coordinates of a composition in such a basis are called *balances* and the compositional vectors making up the basis *balancing elements*.

A *sequential binary partition (SBP)* is a hierarchy of the parts of a composition. In the first order of the hierarchy, all parts are split into two groups. In the following steps, each group is in turn split into two groups. The process continues until all groups have a single part, as illustrated in Table 4.1.

For the kth order partition, it is possible to define the *balance* between the two subgroups formed at that level: if $i_1, i_2, \ldots, i_r$ are the r parts of the first subgroup (coded by $+1$) and $j_1, j_2, \ldots, j_s$ the s parts of the second (coded by -1), the balance is defined as the normalized logratio of the geometric mean of each group of parts:

$$b_k = \sqrt{\frac{rs}{r+s}} \ln \frac{\left(x_{i_1} x_{i_2} \cdots x_{i_r}\right)^{1/r}}{\left(x_{j_1} x_{j_2} \cdots x_{j_s}\right)^{1/s}} = \ln \frac{\left(x_{i_1} x_{i_2} \cdots x_{i_r}\right)^{a_+}}{\left(x_{j_1} x_{j_2} \cdots x_{j_s}\right)^{a_-}}, \qquad (4.9)$$

where

$$a_+ = +\frac{1}{r}\sqrt{\frac{rs}{r+s}}, \qquad a_- = -\frac{1}{s}\sqrt{\frac{rs}{r+s}}. \qquad (4.10)$$

Note that a_+ refers to parts in the numerator, a_- refers to parts in the denominator, and the values of r and s are those corresponding to the kth order partition. The balance is then

$$b_k = \sum_{j=1}^{D} a_{kj} \ln x_j,$$

Table 4.1   Example of sign matrix used to encode a sequential binary partition and build an orthonormal basis. The columns r and s are the number of plus and minus signs in each row respectively. The lower part of the table shows the matrix Ψ of the basis.

Order	x_1	x_2	x_3	x_4	x_5	x_6	r	s
1	+1	+1	−1	−1	+1	+1	4	2
2	+1	−1	0	0	−1	−1	1	3
3	0	+1	0	0	−1	−1	1	2
4	0	0	0	0	+1	−1	1	1
5	0	0	+1	−1	0	0	1	1

Order	x_1	x_2	x_3	x_4	x_5	x_6
1	$\frac{1}{4}\sqrt{\frac{4\cdot2}{4+2}}$	$\frac{1}{4}\sqrt{\frac{4\cdot2}{4+2}}$	$\frac{-1}{2}\sqrt{\frac{4\cdot2}{4+2}}$	$\frac{-1}{2}\sqrt{\frac{4\cdot2}{4+2}}$	$\frac{1}{4}\sqrt{\frac{4\cdot2}{4+2}}$	$\frac{1}{4}\sqrt{\frac{4\cdot2}{4+2}}$
2	$+\frac{\sqrt{3}}{2}$	$-\frac{1}{\sqrt{12}}$	0	0	$-\frac{1}{\sqrt{12}}$	$-\frac{1}{\sqrt{12}}$
3	0	$+\frac{\sqrt{2}}{\sqrt{3}}$	0	0	$-\frac{1}{\sqrt{6}}$	$-\frac{1}{\sqrt{6}}$
4	0	0	0	0	$+\frac{1}{\sqrt{2}}$	$-\frac{1}{\sqrt{2}}$
5	0	0	$+\frac{1}{\sqrt{2}}$	$-\frac{1}{\sqrt{2}}$	0	0

where $a_{kj} = a_+$ if, for the jth part at the kth order partition, the code is +1; $a_{kj} = a_-$ if the code is −1; and $a_{kj} = 0$ if the code is null. Note that the matrix with entries a_{ij} is just the contrast matrix Ψ of Definition 4.3, as shown in the lower part of Table 4.1. It is remarkable that, in an SBP process, changing the sign codes from + into − and vice versa at any step of the partition only causes the change of sign of the associated balance.

The order in which the ilr-coordinates, and particularly balances, are ordered is arbitrary. Hence, the rows of the associated contrast matrix Ψ can be permuted with the only consequence of a permutation of indices of coordinates or balances. For instance, the reversion of the order of an SBP allows to interpret the process of building up the signs code of the SBP as a process of fusion of groups of parts better than a partition. The idea is to start with D groups of parts containing a single part. The first step consists of merging two of these groups into a single one, thus obtaining $D − 1$ groups of parts. In subsequent steps, two groups are again merged until obtaining a single group containing the D parts. This is attained after $D − 1$ fusion steps. In each fusion, one group is marked with +'s and the other one with −'s. This process of fusion is completely equivalent to an SBP up to the reversion of the indices. In this book, both "direct" and "reversed" SBPs are used.

Example 4.6. Egozcue et al. (2003) obtained an orthonormal basis of the simplex using the Gram–Schmidt procedure. It corresponds to the SBP shown in Table 4.2. The main feature of matrix Ψ in Table 4.2 is that, for each level $i = 1, 2, \ldots, D-1$ of the partition, entries can be easily expressed as

$$\psi_{ij} = a_{ij} = \begin{cases} +\sqrt{\dfrac{1}{(D-i)(D-i+1)}}, & j \le D-i; \\[3mm] -\sqrt{\dfrac{D-i}{D-i+1}}, & j = D-i+1; \\[3mm] 0, & \text{otherwise.} \end{cases}$$

This matrix is closely related to Helmert matrices (Lancaster 1965).   ◇

The interpretation of balances mainly relays on their expression, using geometric means in the numerator and denominator as in

$$b = \sqrt{\dfrac{rs}{r+s}} \ln \dfrac{\left(x_1 \ldots x_r\right)^{1/r}}{\left(x_{r+1} \ldots x_D\right)^{1/s}}.$$

The geometric means are central values of the parts in each group; their ratio measures the relative weight of each group; the logarithm provides the appropriate

Table 4.2   Example of sign matrix for $D = 5$, used to encode a sequential binary partition in a standard way. The lower part of the table shows the contrast matrix Ψ of the basis.

Level	x_1	x_2	x_3	x_4	x_5	r	s
1	+1	+1	+1	+1	−1	4	1
2	+1	+1	+1	−1	0	3	1
3	+1	+1	−1	0	0	2	1
4	+1	−1	0	0	0	1	1
1	$+\frac{1}{\sqrt{20}}$	$+\frac{1}{\sqrt{20}}$	$+\frac{1}{\sqrt{20}}$	$+\frac{1}{\sqrt{20}}$	$-\frac{2}{\sqrt{5}}$		
2	$+\frac{1}{\sqrt{12}}$	$+\frac{1}{\sqrt{12}}$	$+\frac{1}{\sqrt{12}}$	$-\frac{\sqrt{3}}{\sqrt{4}}$	0		
3	$+\frac{1}{\sqrt{6}}$	$+\frac{1}{\sqrt{6}}$	$-\frac{\sqrt{2}}{\sqrt{3}}$	0	0		
4	$+\frac{1}{\sqrt{2}}$	$-\frac{1}{\sqrt{2}}$	0	0	0		

scale; and the square root coefficient is a normalizing constant that allows to compare numerically different balances. A positive balance means that, in (geometric) mean, the group of parts in the numerator has more weight in the composition than the group in the denominator (and conversely for negative balances). They behave like the weights in a steelyard, a portable hanging balance used in roman times. This justifies the denomination as *balances*.

Example 4.7 (Election example).
Imagine that in an election, six parties or coalitions have contested. The information collected gives the percentages each party or coalition obtained, that is, a composition in S^6. Divide the parties into two groups, the left and the right wing ones, with four parties in the left wing and two in the right wing. The SBP in Table 4.1 can be used for identifying $\{x_3, x_4\}$ with the right group and $\{x_1, x_2, x_5, x_6\}$ with the left group. If someone is interested in knowing which wing has obtained more votes and in evaluating their relative difference, the balance between the left group versus the right group (first-order partition in Table 4.1) provides this quantitative information:

$$b_1 = \sqrt{\frac{4 \cdot 2}{4 + 2}} \ln \frac{\left(x_1 x_2 x_5 x_6\right)^{1/4}}{\left(x_3 x_4\right)^{1/2}}.$$

The sign of the balance points out which group obtained more votes, and the value gives the size of the difference in log relative scale. Note that in this case, there is no reference to the structure of votes within each wing, that is, it gives only information about the left–right balance. Alternatively, if interest is focussed on the vote within the right group, the fifth-order partition gives the adequate balance, $b_5 = \sqrt{1/2} \ln(x_3/x_4)$. For the left group, made of four parts, three balances (orders 2, 3, and 4 in Table 4.1) are necessary to completely describe the distribution of votes within the left group. An important feature of balances is their geometric orthogonality; it indicates that they provide nonredundant information about the composition. The knowledge of b_1 tells nothing about b_5 or other balances in a geometrical sense, and vice versa. Nevertheless, be aware that geometrical orthogonality does not imply statistical uncorrelation.

One may wonder why balances are used to answer the questions. For instance, the relative difference between the left and right wings could also be described by

$$\ln \frac{x_1 + x_2 + x_5 + x_6}{x_3 + x_4}.$$

But this descriptor is unsuitable for the following reasons. (i) It uses amalgamations of parts; therefore, it is a logratio in S^2. This implies a dramatic change in the considered sample space of six-part compositions, thus lacking consistency for comparisons with other descriptors in S^6. (ii) It does not take into account the number of parties in each group. (iii) It is not an orthogonal projection, as will be seen when studying orthogonal projections in the simplex (Section 4.8).   ◇

Balances can also be defined independently of an SBP and the corresponding orthonormal basis. Roughly speaking, they are coordinates with respect to a special kind of unitary vector of the simplex, called balancing element or balancing composition. The characteristic of balancing elements is that only two groups of parts participate actively in its definition (Egozcue and Pawlowsky-Glahn, 2005).

Definition 4.8 (Balancing element).
A composition $\mathbf{e} = C[e_1, e_2, \ldots, e_D]$ in S^D is a balancing element if its components are divided into three groups, denoted as G_+, G_-, and G_0, containing, respectively, $r > 0$, $s > 0$, and $D - r - s \geq 0$ components, such that the components have values proportional to

$$e_+ = \exp\left(\frac{1}{r}\sqrt{\frac{rs}{r+s}}\right), \quad e_- = \exp\left(-\frac{1}{s}\sqrt{\frac{rs}{r+s}}\right), \quad e_0 = 1$$

in G_+, G_-, G_0, respectively.

A first property of a balancing composition is that it is unitary, $\|\mathbf{e}\|_a = 1$. It can be checked by computing clr(**e**) whose components are $\log e_+$, $\log e_-$, 0 in the groups of components G_+, G_-, G_0, respectively. These components add to 0, as corresponds to a clr-transformed composition, and their squares add to 1, thus $\|\text{clr}(\mathbf{e})\| = 1$.

The definition of balance can be stated as follows.

Definition 4.9 (Balance).
Let $\mathbf{x}$ and $\mathbf{e}$ be, respectively, a composition and a balancing element in S^D. The balance of $\mathbf{x}$ with respect to the balancing element $\mathbf{e}$ is $b = \langle \mathbf{x}, \mathbf{e} \rangle_a$. Using the notation of Definition 4.8, the expression of the balance is

$$b = \sqrt{\frac{rs}{r+s}} \ln \frac{\prod_+ x_i^{1/r}}{\prod_- x_j^{1/s}} = \sum_{G_+} e_+ \ln x_i + \sum_{G_-} e_- \ln x_j. \tag{4.11}$$

From Equation (4.11), it is clear that a balance is a logcontrast, as it can be expressed as a linear combination of the logarithms of the parts in which coefficients add to zero. On the other hand, the projection of $\mathbf{x}$ on $\mathbf{e}$ is $b \odot \mathbf{e}$, that is, b is a normalized coordinate of $\mathbf{x}$ in the direction of $\mathbf{e}$.

4.6 Working on coordinates

The *principle of working on coordinates* was first introduced in Pawlowsky-Glahn (2003) and further developed in Mateu-Figueras et al. (2011). It is a general property of coordinates in real linear vector spaces and is based on the fact that coordinates with respect to an orthonormal basis obey standard rules of operation in real space. If we consider the vector of coordinates $\mathrm{ilr}(\mathbf{x}) = \mathbf{x}^* \in \mathbb{R}^{D-1}$ of a compositional vector $\mathbf{x} \in S^D$ with respect to an arbitrary orthonormal basis, it holds (Property 4.5)

$$\mathrm{ilr}(\mathbf{x} \oplus \mathbf{y}) = \mathrm{ilr}(\mathbf{x}) + \mathrm{ilr}(\mathbf{y}) = \mathbf{x}^* + \mathbf{y}^*, \quad \mathrm{ilr}(\alpha \odot \mathbf{x}) = \alpha \cdot \mathrm{ilr}(\mathbf{x}) = \alpha \cdot \mathbf{x}^*. \quad (4.12)$$

Thus, it is possible to think about perturbation as having the same properties in the simplex as translation has in real space and of the power transformation as having the same properties as multiplication. Furthermore,

$$\mathrm{d}_a(\mathbf{x}, \mathbf{y}) = \mathrm{d}(\mathrm{ilr}(\mathbf{x}), \mathrm{ilr}(\mathbf{y})) = \mathrm{d}(\mathbf{x}^*, \mathbf{y}^*),$$

where d stands for the usual Euclidean distance in real space. This means that, when performing analysis of compositional data, results that could be obtained using compositions and the Aitchison geometry are exactly the same as those obtained using the coordinates of the compositions and using the ordinary Euclidean geometry. This latter possibility reduces computations to ordinary operations in real spaces, thus facilitating to apply standard procedures. The duality of the representation of compositions, in the simplex and in coordinates, introduces a rich framework where both representations can be interpreted to extract conclusions from the analysis (see Figures 4.1–4.4, for illustration). The

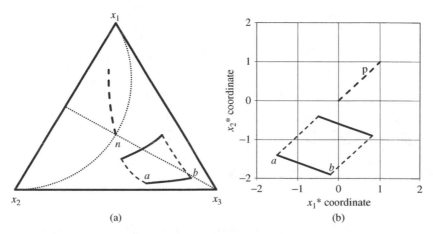

Figure 4.1   Perturbation of a segment in (a) S^3 and (b) coordinates.

price is that the basis selected for representation should be carefully selected for an enhanced interpretation.

Working on coordinates can be done in a blind way, just selecting a default basis, computing the corresponding coordinates, and, when the results in coordinates are obtained, translating the results back into the simplex for interpretation. This is acceptable whenever the method of analysis used is ensured to deliver the same result in any orthonormal basis (i.e., a *rotation-invariant method*). Otherwise, the method would deliver a different result depending on an arbitrary, uncontrolled choice of basis. When a number of compositions are represented by

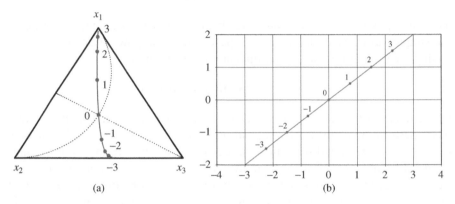

Figure 4.2   Powering of a vector in (a) S^3 and (b) coordinates.

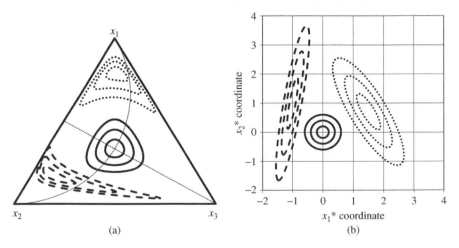

Figure 4.3   Circles and ellipses in (a) S^3 and (b) coordinates.

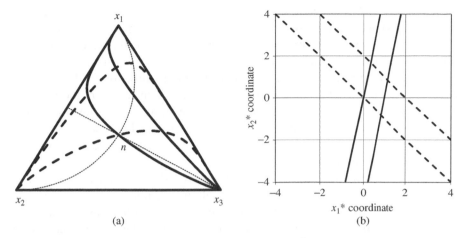

Figure 4.4   Couples of parallel lines in (a) S^3 and (b) coordinates.

their coordinates in a given basis of the simplex, some features can be enhanced while other ones can be hidden. Owing to these facts, the choice of a basis should be preferably oriented to a better interpretation and visualization, thus depending on subjective criteria. The analyst should keep in mind that the appearance of a set of compositions in different coordinates may experience important changes.

One point is essential in the proposed approach: no zero values are allowed, as neither division by zero is admissible nor taking the logarithm of zero. Zeroes in a composition place it at infinity of the simplex when using the Aitchison geometry, and this is what you see when expressing it in coordinates. This is coherent with the possible representation in a ternary diagram, in which the infinity is the border of the triangle. The presence of an actual zero in a composition points out that it is not a composition. Consequently, its coordinates cannot be computed, but their expression in proportions is still possible, thus giving the false impression that it is a composition. Methods on how to approach the problem have been discussed in Section 2.3.

In Chapter 3, Figures 3.3 and 3.4 show some geometric figures represented in the ternary diagram. Coordinate representations allow a better visual comprehension of those figures and justify the names given. Figure 4.1a reproduces the right panel of Figure 3.3 and the representation in coordinates of the same curves is given in Figure 4.1b. A rectilinear segment *ab* (solid line) is perturbed (shifted) by the vector with origin at coordinates $(0, 0)$ and end point at $(1, 1)$ (dashed). Joining the segment *ab* with the shifted one (solid line), a rhomboid is completed with the dashed lines. In coordinates, the parallelogram appears as such, while in the ternary diagram, it is distorted. Moreover, Figure 4.1 shows why perturbation is a shift in the simplex, as it is the sum of vectors in real space. It should be remarked that in Figures 4.1, 4.2, and 4.4, straight lines are easily visualized in coordinates (right panels), which is not the case in the ternary diagram (left panels).

Figure 4.2 shows the effect of powering in both the ternary diagram (Figure 4.2a) and coordinates (Figure 4.2b). A vector in coordinates $\mathbf{x}^*$ from the origin $(0, 0)$ to $(0.832, 0.555)$ (a unitary vector in the direction $(3, 2)$) is labeled with 1. Other points are labeled with $\pm 1, \pm 2, \ldots$, which represent the values of α in the scaled vectors $\alpha \cdot \mathbf{x}^*$, all of them on the straight line shown as a solid line. This line, and the points on it, correspond to the line and points shown in the ternary diagram, illustrating the effect of powering on compositions: if $\mathbf{x} = \mathrm{ilr}^{-1}(\mathbf{x}^*)$, the points correspond to $\alpha \odot \mathbf{x}$.

Figure 4.3a shows the ellipses and circles previously presented in Figure 3.4 in the simplex $\mathcal{S}^3$. The right panel represents all those circles and ellipses in coordinates, where they appear as circles and ellipses in $\mathbb{R}^2$ with the usual shape. The importance of circles and ellipses in statistical analysis relies on the fact that isodensity curves of bivariate normal distributions are circles or ellipses; Figure 4.3a shows the complex shapes that isodensity curves of normal distributions can attain. Note that the shapes of circles centered at the barycentre of the ternary diagram are similar to circles in a plane, thus showing that the Aitchison geometry near the barycentre is similar to the Euclidean geometry on the ternary diagram.

4.7   Additive logratio coordinates (alr)

In Section 4.3, we considered the generating system of the simplex given in Equation (4.2). One of the elements, for example, the last one, can be suppressed to obtain a basis: $\{\mathbf{w}_1, \mathbf{w}_2, \ldots, \mathbf{w}_{D-1}\}$. Then, any composition $\mathbf{x} \in \mathcal{S}^D$ can be written as

$$\mathbf{x} = \bigoplus_{i=1}^{D-1} \ln \frac{x_i}{x_D} \odot \mathbf{w}_i$$

$$= \ln \frac{x_1}{x_D} \odot [e, 1, \ldots, 1, 1] \oplus \ldots \oplus \ln \frac{x_{D-1}}{x_D} \odot [1, 1, \ldots, e, 1].$$

These coordinates correspond to the well-known additive logratio transformation (alr) introduced by Aitchison (1986). It can be defined as follows.

Definition 4.10 (Additive logratio (alr) coordinates).
Let $\mathbf{x} = [x_1, x_2, \ldots, x_D]$ be a composition in $\mathcal{S}^D$ and consider x_D as a reference part. Its alr *transformation into $\mathbb{R}^{D-1}$ is*

$$\mathrm{alr}(\mathbf{x}) = \left[\ln \frac{x_1}{x_D}, \ln \frac{x_2}{x_D}, \ldots, \ln \frac{x_{D-1}}{x_D} \right] = \zeta.$$

The inverse alr *transformation, recovering* $\mathbf{x}$ *from* $\zeta = [\zeta_1, \zeta_2, \ldots, \zeta_{D-1}]$, *is*

$$\mathbf{x} = \mathrm{alr}^{-1}(\zeta) = C[\exp(\zeta_1), \exp(\zeta_2), \ldots, \exp(\zeta_{D-1}), 1].$$

Note that the alr transformation is not symmetric in the components, as the reference part x_D is in the denominator of the component logratios. Another reference part can be chosen, leading to different alr-transformations. For instance, in logistic regression (e.g., Agresti, 1990), it is customary to select x_1 as reference.

As in the case of the ilr transformation, the alr provides $D - 1$ coordinates matching the dimension of S^D. These coordinates, $\zeta_i = \ln(x_i/x_D)$, are simple logratios and easily interpretable, specially when x_D plays a relevant role in the composition. Furthermore, the alr translates perturbation and powering into the sum and multiplication in $\mathbb{R}^{D-1}$, that is,

$$\mathrm{alr}(\alpha_1 \odot \mathbf{x}_1 \oplus \alpha_2 \odot \mathbf{x}_2) = \alpha_1 \cdot \mathrm{alr}(\mathbf{x}_1) + \alpha_2 \cdot \mathrm{alr}(\mathbf{x}_2),$$

where $\mathbf{x}_1, \mathbf{x}_2$ are compositions in S^D, and α_1, α_2 are real constants.

However, alr-coordinates are referred to an oblique basis of the simplex (Egozcue et al., 2011), and the Aitchison inner products or distances cannot be computed on the alr-coordinates in the standard Euclidean way, that is, the alr does not provide an isometry between S^D and $\mathbb{R}^{D-1}$. The alr-coordinates can be used to compute the Aitchison distances after an appropriate matrix transformation (Aitchison, 1986). In fact, clr coefficients and ilr coordinates can be expressed as a linear combination of alr coordinates (Egozcue et al., 2003), and the distances can then be computed as in Properties 4.2(c) or 4.5(d). Exercise 25 illustrates these facts. The alr-transformation is frequent in many applied sciences (see, e.g., Albarède (1995), p. 42), but care should be taken if distances or metric concepts are involved. In general, it is legitimate to apply an analysis method to alr-coordinates if it is *affine equivariant*. A method is affine equivariant if it delivers equal results after translating, rotating, and scaling the data set, meaning that results are translated, rotated, or scaled as well. This means that only vector space operations are involved, but not metric concepts, something that is not always easy to prove.

The name of *additive logratio* transformation, alr, comes from the expression of its inverse (Definition 4.10). Each part of the composition (except the part in the denominator of the alr) is

$$x_i = \frac{\exp(\zeta_i)}{1 + \sum_{j=1}^{D-1} \exp(\zeta_j)},$$

where the denominator is the effect of the closure. The term *additive* comes from the denominator, which is the sum of exponentials, in contrast with other transformations where this feature is multiplicative or hybrid as shown in Aitchison (1982, Table 1). It should be remarked that this terminology is maintained because of historical reasons, and it is not related to the perturbation in the simplex and its properties as an additive group.

4.8   Orthogonal projections

Orthogonal projections are of uppermost importance in geometry and statistics as they are the base of dimension reduction. They are intuitively understood in low-dimension real spaces, but a little harder when dimension exceeds three. When working with compositions, orthogonal projections are reduced to projections in the real space representing compositions in coordinates as preconized by the *the principle of working on coordinates* (Section 4.6). However, compositional data are frequently high dimensional and, consequently, intuition may fail when interpreting and carrying out such projections. This section is intended to explain rules of orthogonal projection in some particular cases frequently appearing in compositional analysis.

Using the notation of orthogonal coordinates of the simplex, a real vector $\mathbf{x}^*$ of dimension $D - 1$ can be orthogonally projected onto a subspace H of dimension $d \leq D - 1$. It consists of expressing $\mathbf{x}^*$ as a sum of two vectors $\mathbf{x}_H^*$, $\mathbf{x}_\perp^*$, the first one in the projection subspace H and the latter in its orthogonal subspace $H_\perp$. The subspace $H_\perp$ is made of the vectors that are orthogonal to any vector in H. It can be proven that, once given H,

$$\mathbf{x}^* = \mathbf{x}_H^* + \mathbf{x}_\perp^*, \quad \mathbf{x}_H^* \in H, \quad \mathbf{x}_\perp^* \in H_\perp, \tag{4.13}$$

is a unique additive decomposition of $\mathbf{x}^*$ for a given H. The vector $\mathbf{x}_H^*$ is then called orthogonal projection of $\mathbf{x}^*$ onto H. Further details can be found in textbooks such as Berberian (1961) or Eaton (1983).

The orthogonal projection can be characterized in several ways. One of the most useful is related to the so-called *least squares* approach. The projection $\mathbf{x}_H^*$ is considered to be an approximation of $\mathbf{x}^*$, and $\mathbf{x}_\perp^*$ a residual or approximation error. The residual $\mathbf{x}_\perp^*$ is orthogonal to any vector in H, that is, for any $\mathbf{x}_1^* \in H$, $\langle \mathbf{x}_\perp^*, \mathbf{x}_1^* \rangle = 0$. Equivalently, the square-norm of the residual $\|\mathbf{x}_\perp^*\|^2$ is minimum, that is, $\mathbf{x}_H^*$ is the least squares approximation of $\mathbf{x}^*$ in H.

Assume that an orthonormal basis $\{\mathbf{h}_1^*, \mathbf{h}_2^*, \ldots, \mathbf{h}_d^*\}$ of the subspace H is given. The projection of $\mathbf{x}^*$ in H is then easily computed as

$$\mathbf{x}_H^* = \sum_{i=1}^{d} \langle \mathbf{x}^*, \mathbf{h}_i^* \rangle \cdot \mathbf{h}_i^*. \tag{4.14}$$

The projection $\mathbf{x}_H^*$ is represented by $D - 1$ coordinates corresponding to $\mathbb{R}^{D-1}$. However, in many circumstances, a representation within the subspace H

is desired, so that the projection is represented using only the coordinates within H. This is easily attained if the basis of $\mathbb{R}^{D-1}$ includes the basis of H. Assume that the orthonormal basis of the whole space $\mathbb{R}^{D-1}$ is $\{\mathbf{h}_1^*, \mathbf{h}_2^*, \ldots, \mathbf{h}_d^*, \mathbf{h}_{d+1}^*, \ldots, \mathbf{h}_{D-1}^*\}$. Then, the expression in coordinates of $\mathbf{x}^*$ is

$$\mathbf{x}^* = \sum_{i=1}^{D-1} \langle \mathbf{x}^*, \mathbf{h}_i^* \rangle \cdot \mathbf{h}_i^*.$$

This reveals that obtaining the projection $\mathbf{x}_H^*$ consists of setting $\langle \mathbf{x}^*, \mathbf{h}_i^* \rangle = 0$ for $i = d+1, d+2, \ldots, D-1$. Consequently, setting to 0 some coordinates is equivalent to an orthogonal projection on the subspace generated by the remaining coordinates. The next step consists of retaining the nonzero coordinates; the result is a vector of reduced dimension d. The following simple example illustrates this technique.

Example 4.11. The vector $\mathbf{x}^* = [1, 2, 3]$ in $\mathbb{R}^3$ has to be projected on the plane H of the first two coordinates. The canonical basis of $\mathbb{R}^3$ is considered, that is, $\mathbf{h}_1^* = [1, 0, 0]$, $\mathbf{h}_2^* = [0, 1, 0]$, $\mathbf{h}_3^* = [0, 0, 1]$. The plane of the first two components is generated by $\mathbf{h}_1^* = [1, 0, 0]$ and $\mathbf{h}_2^* = [0, 1, 0]$. The inner products in Equation (4.14) are computed as follows:

$$\langle \mathbf{x}^*, \mathbf{h}_1^* \rangle = 1 \cdot 1 + 2 \cdot 0 + 3 \cdot 0 = 1, \tag{4.15}$$

$$\langle \mathbf{x}^*, \mathbf{h}_2^* \rangle = 1 \cdot 0 + 2 \cdot 1 + 3 \cdot 0 = 2, \tag{4.16}$$

which, in fact, are the first two coordinates of $\mathbf{x}^*$. The expression of $\mathbf{x}^*$ in $\mathbb{R}^3$ is $\mathbf{x}^* = 1 \cdot [1, 0, 0] + 2 \cdot [0, 1, 0] + 3 \cdot [0, 0, 1]$ and the desired projection is

$$\mathbf{x}_H^* = 1 \cdot [1, 0, 0] + 2 \cdot [0, 1, 0] + 0 \cdot [0, 0, 1] = [1, 2, 0].$$

The projection $\mathbf{x}_H^*$ is obtained by setting to 0 the last coordinate of $\mathbf{x}^*$. To work only in H, the projection $[1, 2, 0]$ can be replaced by $[1, 2] \in \mathbb{R}^2$. The only condition is to remember that the two coordinates correspond to the subspace $H \subseteq \mathbb{R}^3$.

Moreover, the residual vector $\mathbf{h}_\perp^* = [0, 0, 3]$, is orthogonal to H; its square-norm is $\|\mathbf{h}_\perp^*\|^2 = 3^2$, which is the minimum attainable by any vector in H. ◇

These are some basics of orthogonal projections in real space. It can be assumed that all real vectors (with asterisk) are coordinates of compositions in the simplex $\mathcal{S}^D$. Then, these statements are readily translated into compositions, denoted without asterisk. For instance, assume that $\mathbf{x}^*$ are the ilr-coordinates of the composition $\mathbf{x} \in \mathcal{S}^D$. The computation of the orthogonal projection of $\mathbf{x}$ onto a subspace of the simplex $\mathcal{H}$ determined by the orthonormal basis $\{\mathbf{h}_1, \mathbf{h}_2, \ldots, \mathbf{h}_d\}$, where $\mathbf{h}_i^* = \mathrm{ilr}(\mathbf{h}_i)$, is

$$\mathbf{x}_{\mathcal{H}} = \sum_{i=1}^{d} \langle \mathbf{x}, \mathbf{h}_i \rangle_a \odot \mathbf{h}_i, \tag{4.17}$$

which is obtained taking ilr^{-1} in Equation (4.14). This computation of orthogonal projection is now used to deal with some orthogonal projections in the simplex.

Subcompositions as orthogonal projections

Suppose that a composition $\mathbf{x} \in S^D$ is partitioned into two groups of parts: $\mathbf{x}_s$, including the first s parts, and its complementary $\mathbf{x}_c$, made of the remaining $D - s$ parts. The subcomposition corresponding to $\mathbf{x}_s$ is simply $C\mathbf{x}_s$. The goal is to perform an orthogonal projection of $\mathbf{x}$ that preserves the relative information contained in $\mathbf{x}_s$ and, simultaneously, filters out all the relative information involving parts in $\mathbf{x}_c$.

In order to easily study such an orthogonal projection, an orthonormal basis of S^D can be built up using the SBP shown in Table 4.3. The first coordinate x_1^* separates $\mathbf{x}_s$ from $\mathbf{x}_c$; that is, it is the balance between the groups of parts $\mathbf{x}_s$ and $\mathbf{x}_c$. The following $s - 1$ coordinates x_i^*, $i = 2, 3, \ldots, s$ are balances, which only involve parts in $\mathbf{x}_s$. The partition is arbitrary within $\mathbf{x}_s$. Partitions associated with coordinates $x_{d+1}^*, x_{d+2}^*, \ldots, x_{D-1}^*$ is an arbitrary SBP within $\mathbf{x}_c$. They define the $(D - s - 1)$ coordinates (balances) x_i^*, $i = s + 1, s + 2, \ldots, D - 1$, which only involve the parts in $\mathbf{x}_c$. The desired projection is obtained setting to zero the coordinates involving parts of $\mathbf{x}_c$, that is, $x_1^* = 0, x_i^* = 0$ for $i = s + 1, s + 2, \ldots, D - 1$.

Table 4.3   SBP code to illustrate the concept of subcompositions as orthogonal projections. See text for details.

Coordinates	x_1	x_2	$\cdots$	x_s	x_{s+1}	x_{s+2}	$\cdots$	x_D
x_1^*	$+1$	$+1$	$\ldots$	$+1$	-1	-1	$\ldots$	-1
x_2^*	$+1$	$+1$	$\ldots$	-1	0	0	$\ldots$	0
$\vdots$	$\vdots$	$\vdots$	$\ddots$	$\vdots$	$\vdots$	$\vdots$	$\ddots$	$\vdots$
x_s	$+1$	-1	$\ldots$	0	0	0	$\ldots$	0
x_{s+1}	0	0	$\ldots$	0	$+1$	$+1$	$\ldots$	-1
x_{s+2}	0	0	$\ldots$	0	$+1$	$+1$	$\ldots$	0
$\vdots$	$\vdots$	$\vdots$	$\ddots$	$\vdots$	$\vdots$	$\vdots$	$\ddots$	$\vdots$
x_{D-1}^*	0	0	$\ldots$	0	$+1$	-1	$\ldots$	0

The projection subspace H is defined by the balancing elements corresponding to the coordinates x_i^*, $i = 2, 3, \ldots, s$ and the coordinate vector of the projection is

$$\mathbf{x}_H^* = [0, x_2^*, x_3^*, \ldots, x_s^*, 0, 0, \ldots, 0].$$

To recover the projection in the simplex, ilr^{-1} has to be applied to $\mathbf{x}_H^*$. The contrast matrix $\mathbf{\Psi}$ has the structure of zeroes given by the SBP in Table 4.3, which implies that the projected composition in S^D is

$$\mathbf{x}_H = C[x_1, x_2, \ldots, x_s, g_m(\mathbf{x}_s), g_m(\mathbf{x}_s), \ldots, g_m(\mathbf{x}_s)],$$

where there is no reference to parts in $\mathbf{x}_c$. In order to reduce the projection $\mathbf{x}_H^*$ to S^s, the subcomposition of the first s parts is taken, resulting in $C\mathbf{x}_s$. Therefore, the orthogonal projection described here is equivalent to take directly the subcomposition $C\mathbf{x}_s$ from the original composition $\mathbf{x}$.

Example 4.12 (Election example (continued from Example 4.7)).
Consider the six parties that contested an election from Example 4.7. The parties labeled $L = \{x_1, x_2, x_5, x_6\}$ constituted the group of left-wing parties and $R = \{x_3, x_4\}$ the right-wing group of parties. If the information of interest is only the distribution of votes within the left-wing group (L), the balances b_2, b_3, and b_4 (see Example 4.7, Table 4.1) give the relevant information and must be retained; balances b_1 and b_5 inform about the relation between L and R, and within R, respectively. Considering only b_2, b_3, and b_4 corresponds to an orthogonal projection of the composition of percentages of votes $\mathbf{x} \in S^6$ onto the subcomposition including the parts in L. The projected composition in S^6 is

$$\mathbf{x}_{\mathrm{proj}} = C[x_1, x_2, g_m(L), g_m(L), x_4, x_5].$$

As $g_m(L)$ does not involve parts outside the L group, this projection can be handled as the subcomposition $C[x_1, x_2, x_4, x_5]$.　　　　　　　　　　　　　◇

Balances between subcompositions

Continuing with the notation introduced previously, a composition $\mathbf{x} \in S^D$ is partitioned into two groups of parts $\mathbf{x} = [\mathbf{x}_s, \mathbf{x}_c]$ with s and $D - s$ parts, respectively. Now the interest is centered on a relative comparison of $\mathbf{x}_s$, $\mathbf{x}_c$ but removing the information within each group. Again this reduction of information corresponds to an orthogonal projection. It can be carried out using the same SBP used to project into a subcomposition and shown in Table 4.3. In this case, the projection subspace H is one dimensional and generated by the balancing element associated with the first balance x_1^*. The projection thus consists of setting to zero the whole vector of coordinates except the first coordinate:

$$\mathbf{x}_H^* = [x_1^*, 0, 0, \ldots, 0].$$

Taking ilr^{-1}, the resulting projected composition is

$$\mathbf{x}_H = C[\underbrace{g_m(\mathbf{x}_s), g_m(\mathbf{x}_s), \ldots, g_m(\mathbf{x}_s)}_{s \text{ parts}}, \underbrace{g_m(\mathbf{x}_c), g_m(\mathbf{x}_c), \ldots, g_m(\mathbf{x}_c)}_{D-s \text{ parts}}].$$

The removal of the within group relative information is reflected in the substitution of each part in the group by the geometric mean of all its components. The summary of the relative information retained in this projection is the single balance preserved,

$$x_1^* = \sqrt{\frac{s(D-s)}{D}} \; \ln \frac{g_m(\mathbf{x}_s)}{g_m(\mathbf{x}_c)}.$$

A note of caution is convenient here. One can be tempted to summarize the relative information between $\mathbf{x}_s$ and $\mathbf{x}_c$ by means of one of the following logratios

$$r_1 = \ln \frac{g_m(\mathbf{x}_s)}{g_m(\mathbf{x}_c)}, \quad r_2 = \ln \frac{x_1 + x_2 + \cdots + x_s}{x_{s+1} + x_{s+2} + \cdots + x_D}.$$

Although both r_1 and r_2 are informative, they are in some sense biased, as they do not correspond to orthogonal projections. The first logratio, r_1, does not take into account the number of parts present in each group. The second logratio, r_2, uses amalgamation of the two groups of parts, thus reducing the original simplex S^D to a two-part simplex S^2. As commented in Section 2.3.1, amalgamation of parts is problematic when the amalgamated composition is compared with the original one.

Example 4.13 (Election example – continued from Examples 4.7 and 4.12).
Consider the six parties that contested an election from Example 4.7. The parties labeled $L = \{x_1, x_2, x_5, x_6\}$ constituted the group of left-wing parties and $R = \{x_3, x_4\}$ the right-wing group of parties. When someone is only interested in the relative weight of these two groups of parties, the single balance b_1 (Example 4.7) provided the desired information. Considering only b_1 corresponds to an orthogonal projection of the composition of percentages of votes $\mathbf{x} \in S^6$: it filters out all relative information within both groups. The projected composition is

$$\mathbf{x}_{\mathrm{proj}} = C[g_m(L), g_m(L), g_m(R), g_m(R), g_m(L), g_m(L)],$$

where, within each group, each part is substituted by the geometric mean of the parts within the group.   ◊

Projection on a direction

In some cases, orthogonal projection of compositions onto a direction following a given composition is necessary. Alternatively, the projection has to be done on the orthogonal space to that direction. Consider the direction in the simplex defined by a unitary composition $\mathbf{h} \in S^D$. It generates a one-dimensional subspace $\mathcal{H}$ whose elements are $\alpha \odot \mathbf{h}$, $\alpha \in \mathbb{R}$, aligned on a compositional line. The orthogonal projection of $\mathbf{x} \in S^D$ onto $\mathcal{H}$ is

$$\mathbf{x}_{\mathcal{H}} = \langle \mathbf{x}, \mathbf{h} \rangle_a \odot \mathbf{h}.$$

In practice, the computation is carried out by transforming this expression into ilr coordinates or, even simpler, into clr coefficients, thus avoiding the selection of an orthogonal basis. The latter alternative reduces to

$$\mathrm{clr}(\mathbf{x}_{\mathcal{H}}) = \langle \mathrm{clr}(\mathbf{x}), \mathrm{clr}(\mathbf{h}) \rangle \cdot \mathrm{clr}(\mathbf{h}),$$

which is easily computed. The clr^{-1} allows the recovery of $\mathbf{x}_{\mathcal{H}}$.

If the goal is to project $\mathbf{x}$ onto the subspace orthogonal to $\mathbf{h}$, the choice of a basis in this orthogonal subspace is not needed, the desired projection is obtained from $\mathrm{clr}(\mathbf{x}_{\mathcal{H}})$: using Equation (4.17), the required projection is $\mathbf{x}_{\perp} = \mathbf{x} \ominus \mathbf{x}_{\mathcal{H}}$. Transformed into clr, this is

$$\mathrm{clr}(\mathbf{x}_{\perp}) = \mathrm{clr}(\mathbf{x}) - \langle \mathrm{clr}(\mathbf{x}), \mathrm{clr}(\mathbf{h}) \rangle \cdot \mathrm{clr}(\mathbf{h}).$$

For instance, suppose we are interested in the relationships between the parties within the left wing and, consequently, want to remove the information within the right wing. A traditional approach to this is to remove parts x_3 and x_4 and then close the remaining subcomposition. However, this is equivalent to project the composition of six parts orthogonally onto the subspace associated with the left wing, what is easily done by setting $b_5 = 0$. If we do so, the obtained projected composition is

$$\mathbf{x}_{\mathrm{proj}} = C[x_1, x_2, g_m(x_3, x_4), g_m(x_3, x_4), x_5, x_6],$$

with $g_m(x_3, x_4) = (x_3 x_4)^{1/2}$. In words, each part in the right wing is substituted by the geometric mean within the right wing. This composition still has the information on the left–right-wing balance, b_1. If we are also interested in removing it ($b_1 = 0$), the remaining information will be only that within the left-wing subcomposition, which is represented by the orthogonal projection

$$\mathbf{x}_{\mathrm{left}} = C[x_1, x_2, g_m(x_1, x_2, x_5, x_6), g_m(x_1, x_2, x_5, x_6), x_5, x_6],$$

with $g_m(x_1, x_2, x_5, x_6) = (x_1, x_2, x_5, x_6)^{1/4}$. The conclusion is that balances are very useful to project compositions onto special subspaces, just by retaining some balances and making others null.

4.9   Matrix operations in the simplex

Many operations in real spaces are easier expressed in matrix notation. As the simplex is an Euclidean space, matrix notations may also be useful for compositions. However, there is a fundamental difference between the real space and the simplex: in the first case, one can identify an array of scalar values with a vector of the space. But in the case of the simplex, a vector of real constants may not be identifiable with a composition. This produces two kinds of matrix products that are introduced in this section. The first is simply the expression of a perturbation-linear combination of compositions, which appears as a power multiplication of a real vector by a compositional matrix in which rows are compositions. The second one is the expression of a linear transformation in the simplex: a composition in coordinates is transformed by a matrix product, and the result is expressed back as a composition. Each component of the result can be shown to be equivalent to a product of powered components of the original composition, where these powers can be arranged in a square matrix that identifies the transformation. The power matrix implied in this case is not unique, a consequence of the nature of compositions as equivalence classes.

4.9.1   Perturbation-linear combination of compositions

For a row vector of ℓ scalars $\mathbf{a} = [a_1, a_2, \dots, a_\ell]$ and an array of row compositions $\mathbf{X} = (\mathbf{x}_1^\mathsf{T}, \mathbf{x}_2^\mathsf{T}, \dots, \mathbf{x}_\ell^\mathsf{T})^\mathsf{T}$, a perturbation-linear combination of compositions is a matrix product,

$$\mathbf{a} \odot \mathbf{X} = \mathbf{a} \odot \begin{pmatrix} \mathbf{x}_1 \\ \mathbf{x}_2, \\ \dots \\ \mathbf{x}_\ell \end{pmatrix} = [a_1, a_2, \dots, a_\ell] \odot \begin{pmatrix} x_{11} & x_{12} & \cdots & x_{1D} \\ x_{21} & x_{22} & \cdots & x_{2D} \\ \vdots & \vdots & \ddots & \vdots \\ x_{\ell 1} & x_{\ell 2} & \cdots & x_{\ell D} \end{pmatrix}$$

$$= \bigoplus_{i=1}^{\ell} a_i \odot \mathbf{x}_i.$$

The components of this matrix product are

$$\mathbf{a} \odot \mathbf{X} = \mathcal{C} \left[\prod_{i=1}^{\ell} x_{i1}^{a_i}, \prod_{i=1}^{\ell} x_{i2}^{a_i}, \dots, \prod_{i=1}^{\ell} x_{iD}^{a_i} \right].$$

In coordinates, this simplicial matrix product takes the form of a linear combination of the coordinate vectors. In fact,

$$\mathrm{ilr}(\mathbf{a} \odot \mathbf{X}) = \mathrm{ilr}\left(\bigoplus_{i=1}^{\ell} a_i \odot \mathbf{x}_i \right) = \sum_{i=1}^{\ell} a_i \, \mathrm{ilr}(\mathbf{x}_i).$$

Example 4.14. Consider the compositions $\mathbf{x}_1 = [0.1, 0.2, 0.3, 0.4]$ and $\mathbf{x}_2 = [0.7, 0.1, 0.1, 0.1]$, which define the matrix

$$\mathbf{X} = \begin{bmatrix} \mathbf{x}_1 \\ \mathbf{x}_2 \end{bmatrix} = \begin{bmatrix} 0.1, & 0.2, & 0.3, & 0.4 \\ 0.7, & 0.1, & 0.1, & 0.1 \end{bmatrix},$$

and the vector of coefficients $\mathbf{a} = [2, 0.5]$. The perturbation-linear combination of these elements can be expressed in matrix notation as

$$\mathbf{y} = \mathbf{a} \odot \mathbf{X}$$

$$= [2, 0.5] \odot \begin{bmatrix} 0.1, & 0.2, & 0.3, & 0.4 \\ 0.7, & 0.1, & 0.1, & 0.1 \end{bmatrix}$$

$$= \mathcal{C}\left[0.1^2 \sqrt{0.7}, 0.2^2 \sqrt{0.1}, 0.3^2 \sqrt{0.1}, 0.4^2 \sqrt{0.1} \right]$$

$$= [0.084, 0.126, 0.284, 0.506],$$

using the fact that $a^{0.5} = \sqrt{a}$. The reader may try to obtain $\mathbf{y}$ by using the equivalence in ilr coordinates, for instance, using the contrast matrix

$$\boldsymbol{\Psi} = \begin{pmatrix} +\frac{1}{2} & +\frac{1}{2} & -\frac{1}{2} & -\frac{1}{2} \\ 0 & 0 & +\frac{1}{\sqrt{2}} & -\frac{1}{\sqrt{2}} \\ +\frac{1}{\sqrt{2}} & -\frac{1}{\sqrt{2}} & 0 & 0 \end{pmatrix}.$$

◇

Example 4.15. A composition in S^D can be expressed as a perturbation-linear combination of the elements of a given basis $\mathbf{e}_i$, $i = 1, 2, \dots, D-1$, as in Equation (4.6). Consider the $(D-1, D)$-matrix $\mathbf{E} = (\mathbf{e}_1^\mathsf{T}, \mathbf{e}_2^\mathsf{T}, \dots, \mathbf{e}_{D-1}^\mathsf{T})^\mathsf{T}$ and the vector of coordinates in this basis, $\mathbf{x}^* = \mathrm{ilr}(\mathbf{x})$. Equation (4.6) can be rewritten as

$$\mathbf{x} = \mathbf{x}^* \odot \mathbf{E}.$$

◇

4.9.2 Linear transformations of S^D: endomorphisms

Consider a row vector of coordinates $\mathbf{x}^* \in \mathbb{R}^{D-1}$ and a general $(D-1, D-1)$-matrix $\mathbf{A}^*$. In the real-space setting, $\mathbf{y}^* = \mathbf{x}^* \mathbf{A}^*$ represents an endomorphism, a linear transformation of $\mathbb{R}^{D-1}$ onto itself. Given the isometry between the real space of coordinates and the simplex, the endomorphism represented by the matrix $\mathbf{A}^*$ can be translated into an equivalent expression in the simplex. Taking ilr^{-1} and using Equation (4.8),

$$\mathbf{y} = \mathrm{ilr}^{-1}(\mathbf{y}^*) = \mathcal{C}(\exp\,[\mathbf{x}^* \mathbf{A}^* \boldsymbol{\Psi}]) = \mathcal{C}(\exp\,[\mathrm{clr}(\mathbf{x}) \boldsymbol{\Psi}^\mathsf{T} \mathbf{A}^* \boldsymbol{\Psi}]), \qquad (4.18)$$

where $\boldsymbol{\Psi}$ is the contrast matrix of the selected basis (Definition 4.3), and the right-most member has been obtained by applying Equation (4.7) to $\mathbf{x}^*$. The (D, D)-matrix $\mathbf{A} = \boldsymbol{\Psi}^\top \mathbf{A}^* \boldsymbol{\Psi}$ has entries

$$a_{ij} = \sum_{k=1}^{D-1} \sum_{m=1}^{D-1} \psi_{ki} \psi_{mj} a_{km}^*, \quad i, j = 1, 2, \dots, D.$$

Substituting $\mathrm{clr}(\mathbf{x})$ by its expression as a function of the logarithms of parts, the composition $\mathbf{y}$ is

$$\mathbf{y} = C \left[\prod_{j=1}^{D} x_j^{a_{j1}}, \prod_{j=1}^{D} x_j^{a_{j2}}, \dots, \prod_{j=1}^{D} x_j^{a_{jD}} \right].$$

Taking into account that products and powers match the definitions of $\oplus$ and $\odot$, and defining $\mathbf{A} = \boldsymbol{\Psi}^\top \mathbf{A}^* \boldsymbol{\Psi}$, this can be written as

$$\mathbf{y} = \mathbf{x} \boxdot \mathbf{A} = \mathbf{x} \boxdot (\boldsymbol{\Psi}^\top \mathbf{A}^* \boldsymbol{\Psi}), \tag{4.19}$$

where $\boxdot$ stands for the matrix product, representing an endomorphism in the simplex. Note that $\boxdot$ cannot be commutative, as the first term $\mathbf{x}$ is a composition and the second term is a matrix of real constants. This matrix product in the simplex should not be mixed up with the one defined between a vector of scalars and a matrix of compositions, denoted by $\odot$.

An important conclusion is that endomorphisms in the simplex are represented by matrices with a peculiar structure given by $\mathbf{A} = \boldsymbol{\Psi}^\top \mathbf{A}^* \boldsymbol{\Psi}$, which have some remarkable properties:

Property 4.16 Characterization of an endomorphism matrix $\mathbf{A}$:

(a) $\mathbf{A}$ is a (D, D) real matrix;

(b) *zero sum property:* each row and each column of $\mathbf{A}$ adds to 0;

(c) $\mathrm{rank}(\mathbf{A}) = \mathrm{rank}(\mathbf{A}^*)$; particularly, when $\mathbf{A}^*$ is full rank, $\mathrm{rank}(\mathbf{A}) = D - 1$;

(d) the identity endomorphism corresponds to $\mathbf{A}^* = \mathbf{I}_{D-1}$, the identity in $\mathbb{R}^{D-1}$, and to $\mathbf{A} = \boldsymbol{\Psi}^\top \boldsymbol{\Psi} = \mathbf{I}_D - (1/D)\mathbf{1}_D^\top \mathbf{1}_D$, where $\mathbf{I}_D$ is the identity (D, D)-matrix and $\mathbf{1}_D$ is a D-row vector of ones.

The matrix $\mathbf{A}^*$ can be recovered from $\mathbf{A}$ as $\mathbf{A}^* = \boldsymbol{\Psi} \mathbf{A} \boldsymbol{\Psi}^\top$. If $\mathbf{A}$ satisfies the zero sum property, then $\mathbf{A}$ is the only matrix that generates $\mathbf{A}^*$ (Egozcue et al., 2011). However, if the zero sum property is removed, one obtains a transformation that can be expressed as a matrix multiplication, but that is nonlinear (see the next section). Endomorphisms are extensively used and studied in Section 9.5.3, where more properties of these transformations are given. Other algebraic properties of the matrix representations $\mathbf{A}$ and $\mathbf{A}^*$ are given in Section A.3.

Example 4.17. Consider the following matrix in $\mathbb{R}^2$,

$$\mathbf{A}^* = \begin{pmatrix} \frac{1}{2} & \frac{\sqrt{3}}{2} \\ -\frac{\sqrt{3}}{2} & \frac{1}{2} \end{pmatrix},$$

which represents a clockwise rotation of $30°$. Consider this transformation as applied to the vector of orthonormal coordinates defined by the contrast matrix

$$\boldsymbol{\Psi} = \begin{pmatrix} +\frac{1}{\sqrt{6}} & +\frac{1}{\sqrt{6}} & -\frac{2}{\sqrt{6}} \\ +\frac{1}{\sqrt{2}} & -\frac{1}{\sqrt{2}} & 0 \end{pmatrix}.$$

The basis-independent matrix representation of this endomorphism is then

$$\mathbf{A} = \boldsymbol{\Psi}^\mathsf{T}\mathbf{A}^*\boldsymbol{\Psi} = \begin{pmatrix} 1/3 & 1/3 & -2/3 \\ -2/3 & 1/3 & 1/3 \\ 1/3 & -2/3 & 1/3 \end{pmatrix}.$$

Thus we can write this $30°$-rotation of a compositional vector as

$$\mathbf{y} = \mathbf{x} \boxdot \mathbf{A} = C\left[\sqrt[3]{\frac{x_1 x_3}{x_2^2}}, \sqrt[3]{\frac{x_1 x_2}{x_3^2}}, \sqrt[3]{\frac{x_2 x_3}{x_1^2}} \right].$$

$\diamond$

4.9.3 Other matrix transformations on $\mathcal{S}^D$: nonlinear transformations

In general, $\mathbf{A}$ is not the only matrix corresponding to $\mathbf{A}^*$. Consider the following (D, D)-matrix

$$\mathbf{B} = \mathbf{A} + \sum_{i=1}^{D} c_i (\mathbf{e}_i)^\mathsf{T} \mathbf{1}_D + \sum_{j=1}^{D} d_j \mathbf{1}_D^\mathsf{T} \mathbf{e}_j,$$

where $\mathbf{A}$ satisfies the zero sum property, $\mathbf{e}_i = [0, 0, \ldots, 1, \ldots, 0, 0]$ is the ith row vector in the canonical basis of $\mathbb{R}^D$, and c_i, d_j are arbitrary constants. Each additive term in this expression adds a constant row or column, being the remaining entries null. A straightforward development proves that $\mathbf{A}^* = \boldsymbol{\Psi}\mathbf{B}\boldsymbol{\Psi}^\top = \boldsymbol{\Psi}\mathbf{A}\boldsymbol{\Psi}^\top$. To obtain $\mathbf{A}$ from $\mathbf{B}$, first compute $\mathbf{A}^* = \boldsymbol{\Psi}\mathbf{B}\boldsymbol{\Psi}^\top$ and then compute

$$\mathbf{A} = \boldsymbol{\Psi}^\top\mathbf{B}^*\boldsymbol{\Psi} = \boldsymbol{\Psi}^\top\boldsymbol{\Psi}\mathbf{B}\boldsymbol{\Psi}^\top\boldsymbol{\Psi} = \left(\mathbf{I}_D - \left(\frac{1}{D}\right)\mathbf{1}_D^\top\mathbf{1}_D\right)\mathbf{B}\left(\mathbf{I}_D - \left(\frac{1}{D}\right)\mathbf{1}_D^\top\mathbf{1}_D\right),$$

where the second member is the required computation and the third member states that it is equivalent to add constant rows and columns to $\mathbf{A}$.

When a (D, D)-matrix $\mathbf{B}$ does not satisfy the zero sum property, the matrix product $\mathbf{y} = \mathbf{x} \boxdot \mathbf{B}$ does not represent a linear transformation in the simplex,

$$(\mathbf{x}_1 \boxdot \mathbf{B}) \oplus ((\alpha \odot \mathbf{x}_2) \boxdot \mathbf{B}) \neq (\mathbf{x}_1 \oplus (\alpha \odot \mathbf{x}_2)) \boxdot \mathbf{B}, \quad \mathbf{x}_1, \mathbf{x}_2 \in S^D, \ \alpha \in \mathbb{R}.$$

The following example illustrates what happens in this case.

Example 4.18. Consider a $(3, 3)$-matrix $\mathbf{B}$ given by

$$\mathbf{B} = \begin{pmatrix} 0 & 0 & 1 \\ 0 & 2 & 0 \\ -3 & 0 & 0 \end{pmatrix},$$

which does not satisfy the zero sum property. To show that $\mathbf{B}$ is a matrix corresponding to a nonlinear transformation, some points on a line in the simplex are going to be transformed using $\mathbf{B}$. This is best done in coordinates. For instance, consider the contrast matrix $\boldsymbol{\Psi}$ and the corresponding coordinates used in Example 4.17. In Figure 4.5, a set of 20 points, $\mathbf{x}_i$, has been regularly distributed on a segment of the line $x_1^* = x_2^*$ spanning $(-10, 10)$ on the x_1^* coordinate (triangle markers). Then, the images $\mathbf{x}_i \boxdot \mathbf{B}$ have been computed, transformed into coordinates, and plotted (empty circles). As the points x_i^* are aligned, the image points would also be aligned if $\mathbf{x}_i \boxdot \mathbf{B}$ were a linear transformation. This is not the case with the images $\mathbf{x}_i \boxdot \mathbf{B}$, which describe a parabola-like curve.

The nonlinearity of the transformation with the matrix $\mathbf{B}$ is not associated with the particular values in $\mathbf{B}$, but with the fact that the zero sum property is not honored. This is shown with a linear transformation derived from $\mathbf{B}$. Consider the matrix

$$\mathbf{A}^* = \boldsymbol{\Psi}\mathbf{B}\boldsymbol{\Psi}^\top = \begin{pmatrix} 1. & -1.1547 \\ 1.1547 & 1. \end{pmatrix}.$$

The matrix $\mathbf{A}^*$ corresponds to a linear transformation of coordinates. The matrix $\mathbf{A}^*$ is in fact an orthogonal transformation (columns are orthogonal as real vectors). It can be associated with a linear transformation in the simplex which

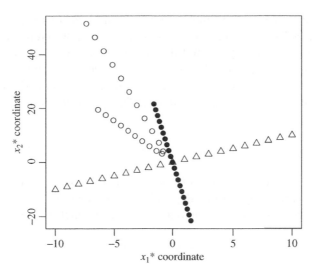

Figure 4.5   A set of 20 compositions **x** *is represented as points in the coordinate space aligned on a segment (triangles). The compositions are nonlinearly transformed as* **x** ⊡ **B** *(hollow circles) and linearly transformed as* **x** ⊡ **A** *(filled circles).*

matrix is

$$\mathbf{A} = \boldsymbol{\Psi}^{\top} \mathbf{A}^{*} \boldsymbol{\Psi} = \frac{1}{3} \begin{pmatrix} 2 & -3 & 1 \\ 1 & 2 & -3 \\ -3 & 1 & 2 \end{pmatrix}.$$

Note that **A** is the result of removing the mean value of rows and columns from **B**, thus forcing the zero sum property. The images of the points $\mathbf{x}_i$ after the transformation $\mathbf{x}_i$ ⊡ **A** are also shown in Figure 4.5 (filled circles). These points, plotted in coordinates, are aligned as expected for a linear transformation and appear to follow a line orthogonal to the original alignment.   ◇

4.10   Coordinates leading to alternative Euclidean structures

Along this chapter, logratio transformations have been used to define coordinates on the simplex. Particularly, isometric logratio transformations (ilr) provide Cartesian coordinates corresponding to the proposed Euclidean structure of the simplex known as the Aitchison geometry. However, this Euclidean structure is not unique, and there are infinitely many alternatives. In fact, any one-to-one mapping $\phi : S^D \to \mathbb{R}^{D-1}$ provides a new Euclidean structure, defined by the operations of the vector space and the corresponding Euclidean metrics (inner product, norm and distance). The vector space operations, now subscripted by ϕ, are readily defined as follows: let $\mathbf{x}$, $\mathbf{y}$ be compositions in S^D, and $\mathbf{u} = \phi(\mathbf{x})$, $\mathbf{v} = \phi(\mathbf{y})$. Consider also a real constant α. The ϕ-perturbation and ϕ-powering can be defined as

$$\mathbf{x} \oplus_\phi \mathbf{y} = \phi^{-1}(\mathbf{u} + \mathbf{v}), \quad \alpha \odot_\phi \mathbf{x} = \phi^{-1}(\alpha \cdot \mathbf{u}), \quad \alpha \in \mathbb{R}, \tag{4.20}$$

where $\mathbf{u}$ and $\mathbf{v}$ play the role of coordinates of $\mathbf{x}$ and $\mathbf{y}$, respectively. The simplex S^D endowed with the operations $\oplus_\phi$ and $\odot_\phi$ is routinely proven to be a $(D-1)$-dimensional vector space over the real space $\mathbb{R}$. The metrics can be defined in a compatible way as

$$\langle \mathbf{x}, \mathbf{y} \rangle_\phi = \langle \mathbf{u}, \mathbf{v} \rangle,$$

and, accordingly,

$$d_\phi(\mathbf{x}, \mathbf{y}) = d(\mathbf{u}, \mathbf{v}), \quad \|\mathbf{x}\|_\phi = \|\mathbf{u}\|,$$

where $\langle \cdot, \cdot \rangle$, $d(\cdot, \cdot)$ and $\| \cdot \|$ denote the ordinary Euclidean inner product, distance, and norm in $\mathbb{R}^{D-1}$, thus completing the new Euclidean structure of S^D. The way of defining the inner product indicates that the ϕ-coordinates of $\mathbf{u} = \phi(\mathbf{x})$ are orthonormal. The advantage of working in orthonormal coordinates of a Euclidean space $\mathcal{E}$ over the real numbers $\mathbb{R}$ is that any property proven in coordinates holds for the space $\mathcal{E}$, which is, as stated in Section 4.6, the *principle of working on coordinates* (Pawlowsky-Glahn, 2003; Mateu-Figueras et al., 2011).

The question arising here is the interpretability and relevance of the ϕ-operations and metrics. The Aitchison perturbation, powering and metrics, have been interpreted in Chapter 3 and proven to fulfill the requirements of a compositional analysis stated as the principles described in Chapter 1. The ϕ-geometries and their corresponding coordinates would be sensible approaches if the ϕ-operations and metrics are shown to fulfill relevant general principles and if they are interpretable. Apart from the Aitchison geometry, the authors are not aware of any ϕ-mapping which is in agreement with the principles of compositional analysis. Nevertheless, some of them could have properties that make them useful for some specific application. The next examples show shortcomings of some simple ϕ-mappings. See also Exercise 27.

Example 4.19 (The alr *as an orthonormal system of coordinates).*
If the mapping ϕ is taken to be identical to the additive logratio transformation alr (Definition 4.10), the ϕ-perturbation and ϕ-powering are identical to the

perturbation and powering in the Aitchison geometry. However, the ϕ-metrics is different. For instance, the square ϕ-distance between two compositions $\mathbf{x}$ and $\mathbf{y}$ is

$$d_{\text{alr}}^2(\mathbf{x}, \mathbf{y}) = \sum_{i=1}^{D-1} \left(\ln \frac{x_i}{x_D} - \ln \frac{y_i}{y_D} \right)^2,$$

which depends on the choice of the part in the denominator, thus violating the permutation invariance principle. ◇

Example 4.20 (Symmetrised ratio transformation).
Let $\mathbf{x}$ be a composition in S^3 and define the mapping $\phi(\mathbf{x}) : S^3 \rightarrow \mathbb{R}^2$, such that

$$\phi(\mathbf{x}) = (u_1, u_2) = \left(\frac{x_1}{x_3} - \frac{x_3}{x_1}, \frac{x_2}{x_3} - \frac{x_3}{x_2} \right).$$

The mapping is one to one because $\mathbf{x}$ can be uniquely recovered from (u_1, u_2). In fact,

$$x_1 = \frac{1}{K(u_1, u_2)} \left(u_1 + \sqrt{u_1^2 + 4} \right), \quad x_2 = \frac{1}{K(u_1, u_2)} \left(u_2 + \sqrt{u_2^2 + 4} \right),$$

$$x_3 = \frac{2}{K(u_1, u_2)}, \quad K(u_1, u_2) = 2 + \left(u_1 + \sqrt{u_1^2 + 4} \right) + \left(u_2 + \sqrt{u_2^2 + 4} \right).$$

According to Equation (4.20), the ϕ-coordinates of $\mathbf{z} = \mathbf{x} \oplus_\phi \mathbf{y}$ are $w_1 = u_1 + v_1$ and $w_2 = u_2 + v_2$, thus leading to a hardly understandable ϕ-perturbation. The first component of such a perturbation is

$$z_1 = \frac{\left(u_1 + v_1 + \sqrt{(u_1 + v_1)^2 + 4} \right)}{2 + \left(u_1 + v_1 + \sqrt{(u_1 + v_1)^2 + 4} \right) + \left(u_2 + v_2 + \sqrt{(u_2 + v_2)^2 + 4} \right)},$$

where

$$u_i + v_i = \frac{x_1}{x_3} + \frac{y_1}{y_3} - \frac{x_3}{x_1} - \frac{y_3}{y_1}, \quad i = 1, 2,$$

and similarly for the second and third components. This ϕ-perturbation violates the principle of permutation invariance and fails at interpretability requirements. However, by construction, it provides a commutative group operation in S^3. ◇

4.11 Exercises

Exercise 14. Consider the data set given in Table 2.1. Compute the clr coefficients defined in Equation (4.3). Verify that the sum of the transformed components equals zero.

Exercise 15. Using the sign matrix of Table 4.1 and Equation (4.10), compute the coefficients of the corresponding contrast matrix, a 5×6 matrix. Which are the balancing elements in the corresponding basis?

Exercise 16. The two rows of the following matrix are six-part compositions in percentage:

$$\begin{pmatrix} \mathbf{x} \\ \mathbf{y} \end{pmatrix} = \begin{pmatrix} 3.74 & 9.35 & 16.82 & 18.69 & 23.36 & 28.04 \\ 9.35 & 28.04 & 16.82 & 3.74 & 18.69 & 23.36 \end{pmatrix}.$$

Using the binary partition of Table 4.1 and Equation (4.9), compute their five balances. Compute their Aitchison squared-norms and inner product. Which is the angle between the two compositions?

Exercise 17. Consider the logratios $c_1 = \ln x_1/x_3$ and $c_2 = \ln x_2/x_3$ in a simplex S^3. They are coordinates when using the alr transformation. Find two unitary vectors $\mathbf{e}_1$ and $\mathbf{e}_2$ such that $\langle \mathbf{x}, \mathbf{e}_i \rangle_a = c_i$, $i = 1, 2$. Compute the inner product $\langle \mathbf{e}_1, \mathbf{e}_2 \rangle_a$, and determine the angle between them. Does the result change if the considered simplex is S^7?

Exercise 18. When computing the clr of a composition $\mathbf{x} \in S^D$, a clr coefficient is $\xi_i = \ln (x_i/g_m(\mathbf{x}))$. It can be considered as a balance between two groups of parts. Which are the parts and which is the corresponding balancing element?

Exercise 19. Six parties have contested an election. In five districts, they have obtained the votes in Table 4.4. Parties are divided into left (L) and right (R) wing parties. Is there some relationship between the $L-R$ balance and the relative votes of $R1-R2$? Select an adequate SBP to analyze this question and obtain the corresponding balance coordinates. Find the correlation matrix of the balances and give an interpretation to the maximum correlated two balances. Compute the

Table 4.4   Votes obtained by six parties in five districts

District	L1	L2	R1	R2	L3	L4
d1	10	223	534	23	154	161
d2	43	154	338	43	120	123
d3	3	78	29	702	265	110
d4	5	107	58	598	123	92
d5	17	91	112	487	80	90

distances between the five districts. Which are the two districts with the maximum and minimum interdistance? Are you able to distinguish some cluster of districts?

Exercise 20. Consider the data set given in Table 2.1. Check the data for zeros. Apply the alr transformation to compositions with no zeros. Plot the transformed data in $\mathbb{R}^2$.

Exercise 21. Consider the data set given in Table 2.1 and take the components in a different order. Apply the alr transformation to compositions with no zeros. Plot the transformed data in $\mathbb{R}^2$. Compare the result with those obtained in Exercise 20.

Exercise 22. Consider the data set given in Table 2.1. Apply the ilr transformation to compositions with no zeros. Plot the transformed data in $\mathbb{R}^2$. Compare the result with the scatterplot obtained in Exercises 20 and 21 using the alr transformation.

Exercise 23. Compute the alr and ilr coordinates, as well as the clr coefficients of the six-part composition

$$[x_1, x_2, x_3, x_4, x_5, x_6] = [3.74, 9.35, 16.82, 18.69, 23.36, 28.04]\%.$$

Exercise 24. Consider the six-part composition of the preceding exercise. Using the binary partition of Table 4.1 and Equation (4.9), compute its five balances. Compare with the results of the preceding exercise.

Exercise 25. Show that the vectors of an alr basis of a four-part composition are not orthogonal.

Exercise 26. Consider two $(D-1, D)$-contrast matrices Ψ_0, Ψ_1 defining ilr-transformations, $ilr_0(\cdot)$, $ilr_1(\cdot)$. Prove

(a) if $\mathbf{x} \in S^D$, then $ilr_i(\mathbf{x})\Psi_i = clr(\mathbf{x})$, $i = 1, 2$.

(b) $[\Psi_1\Psi_2^\mathsf{T}]^{-1} = \Psi_2\Psi_1^\mathsf{T}$.

Exercise 27. Consider a composition $\mathbf{x} = [x_1, x_2, x_3]$ in S^3 and the two coordinates $w_1 = \ln (x_1/(x_2 + x_3))$, $w_2 = \ln (x_2/x_3)$, which constitute a transformation of S^3 onto $\mathbb{R}^2$, known as the *multiplicative logratio* transformation (Aitchison 1982; 1986).

(a) Check that this transformation is one to one by finding the inverse transformation, that is, expressing $\mathbf{x} = [x_1, x_2, x_3]$ as a function of $[w_1, w_2]$.

(b) Assume that $\mathbf{x}$, $\mathbf{y}$ are compositions in $\mathcal{S}^3$ and their coordinates are $[w_1, w_2]$ and $[z_1, z_2]$, respectively. Find the composition $\mathbf{x} \oplus_m \mathbf{y}$ in which coordinates are $[w_1 + z_1, w_2 + z_2]$. Is the perturbation $\oplus_m$ easily interpretable?

(c) Consider the straight lines in $\mathbb{R}^2$ given by $w_1 = -2$, $w_1 = 0$, $w_1 = 2$, $w_2 = -2$, $w_2 = 0$, $w_2 = 2$, and the circle $w_1^2 + w_2^2 = 2$. Plot them in a ternary diagram and ilr-coordinates.

5

Exploratory data analysis

5.1 General remarks

In this chapter, the first steps in any analysis of a compositional data set are addressed. The set is represented as a matrix $\mathbf{X} = [x_{ij}]$ with n rows (observed compositions) and D columns (parts). An exploratory analysis includes the following steps:

1. computing descriptive statistics, that is, the center and variation matrix of a data set, as well as its total variability;

2. looking at the biplot of the data set to discover patterns;

3. plotting patterns in ternary diagrams of subcompositions, possibly centered to enhance visualization;

4. defining an appropriate representation in orthonormal coordinates and computing the corresponding coordinates; and

5. computing classical summary statistics of the coordinates and representing the results in a balance-dendrogram.

In general, the last two steps will be based on a particular sequential binary partition, defined either a priori or as a result of the insights provided by the first three steps.

Modeling and Analysis of Compositional Data, First Edition.
Vera Pawlowsky-Glahn, Juan José Egozcue and Raimon Tolosana-Delgado.
© 2015 John Wiley & Sons, Ltd. Published 2015 by John Wiley & Sons, Ltd.

Before starting, some general considerations need to be made. The first step in a statistical analysis is to check the data set for errors. It can be done using standard procedures, for example, using the minimum and maximum of each part to check whether the values are within an admissible range. Another step is to check the data set for outliers, a point that is outside the scope of this book. See Barceló et al. (1994, 1996), Filzmoser and Hron (2008), Filzmoser et al. (2009), or Filzmoser and Hron (2011), for methods that take outliers into account. Recall that outliers can be considered as such only with respect to a given distribution and that outliers are not erroneous data!

5.2   Sample center, total variance, and variation matrix

Standard descriptive statistics are not very informative in the case of compositional data. In particular, the arithmetic mean and the variance or standard deviation of individual parts do not fit with the Aitchison geometry as value of central tendency and measures of dispersion. The skeptic reader might convince himself/herself by doing Exercise 28.These statistics were defined in the framework of Euclidean geometry in real space, which is not a sensible geometry for compositional data. Therefore, it is necessary to introduce alternatives. They are found in the concepts of *center* (Aitchison, 1997), *variation matrix*, and *total variance* (Aitchison, 1986). Along this chapter, estimators of center, variance and total variance are denoted by var, cen and totvar.

Definition 5.1 (Sample centre).
A value of central tendency of a compositional sample is the closed geometric mean. It is called center. *For a data set of size n, it is defined as*

$$\text{cen}(\mathbf{X}) = \hat{\mathbf{g}} = C\left[\hat{g}_1, \hat{g}_2, \dots, \hat{g}_D\right],$$

with $\hat{g}_j = \left(\prod_{i=1}^n x_{ij}\right)^{1/n}, j = 1, 2, \dots, D.$

Note that in the definition of center of a data set, the geometric mean is considered column-wise (i.e., by parts), while in the clr transformation, given in Equation (4.3), the geometric mean was considered row-wise (i.e., by samples).

Definition 5.2 (Variation matrix).
Dispersion in a compositional data set can be described either by the variation matrix, *originally defined by Aitchison (1986) as*

$$\mathbf{T} = \begin{pmatrix} t_{11} & t_{12} & \cdots & t_{1D} \\ t_{21} & t_{22} & \cdots & t_{2D} \\ \vdots & \vdots & \ddots & \vdots \\ t_{D1} & t_{D2} & \cdots & t_{DD} \end{pmatrix}, \quad t_{ij} = \text{var}\left(\ln \frac{x_i}{x_j}\right),$$

or by the normalized variation matrix

$$
\mathbf{T}^* = \begin{pmatrix} t_{11}^* & t_{12}^* & \cdots & t_{1D}^* \\ t_{21}^* & t_{22}^* & \cdots & t_{2D}^* \\ \vdots & \vdots & \ddots & \vdots \\ t_{D1}^* & t_{D2}^* & \cdots & t_{DD}^* \end{pmatrix}, \quad t_{ij}^* = \mathrm{var}\left(\frac{1}{\sqrt{2}}\ln\frac{x_i}{x_j}\right).
$$

As can be seen, t_{ij} stands for the usual variance of the logratio of parts i and j, while t_{ij}^* stands for the usual variance of the balance of parts i and j (Equation 4.9).

Two-part logratios are balances once normalized by the factor $1/\sqrt{2}$. Each of these variances can be estimated with the unbiased or the maximum likelihood estimators. Under normality of the logratios, the maximum likelihood estimator of the variation matrix elements is

$$
\hat{t}_{ij} = \frac{1}{n}\sum_{k=1}^{n}\left(\ln\frac{x_{ki}}{x_{kj}} - \ln\frac{\hat{g}_i}{\hat{g}_j}\right)^2,
$$

with $\hat{g}_i$ and $\hat{g}_j$ the ith and jth parts of the geometric center of Definition 5.1. Note that

$$
t_{ij}^* = \mathrm{var}\left(\frac{1}{\sqrt{2}}\ln\frac{x_i}{x_j}\right) = \frac{1}{2}t_{ij},
$$

and thus $\mathbf{T}^* = \frac{1}{2}\mathbf{T}$. The same holds for their estimators. Normalized variations have squared Aitchison distance units (see Figure 3.2 for a representation of Aitchison distance units).

Definition 5.3 (Sample total variance).
A measure of global dispersion of a compositional sample is the total variance, *given by*

$$
\mathrm{totvar}\,[\mathbf{X}] = \frac{1}{2D}\sum_{i,j=1}^{D}\mathrm{var}\left(\ln\frac{x_i}{x_j}\right) = \frac{1}{2D}\sum_{i,j=1}^{D}t_{ij} = \frac{1}{n}\sum_{k=1}^{n}d_a^2(\mathbf{x}_k,\hat{\mathbf{g}}).
$$

The last expression, as an average squared Aitchison distance, motivates calling it also *metric variance* (Pawlowsky-Glahn and Egozcue, 2001). By definition, $\mathbf{T}$ and $\mathbf{T}^*$ are symmetric and their diagonals contain only zeros. Furthermore, neither the total variance nor any single entry in both variation matrices depend on the constant κ associated with the sample space S^D, as constants cancel out when taking ratios. Consequently, the closure of the data has no effect on the estimates of these variability measures.

These statistics have further connections. From their definition, it is clear that the total variance summarizes the variation matrix in a single quantity, both in the normalized and nonnormalized version. Note as well that it is possible (and natural) to define a total variance, because all parts in a composition share a common scale. It is by no means so straightforward to define a total variance for a pressure–temperature random vector, for instance. Conversely, any of the two variation matrices admit an interpretation in terms of shares of this variance among the parts: both explain how the total variance is split into contributions of each logratio.

These considerations offer a way to interpret the elements of the variation matrix relative to this total variance: the idea is to normalize the elements of the variation matrix by an average of the total variance, that is, by the value each element of the variation matrix would have if they were all equal,

$$\tau_{ij} = \frac{t_{ij}}{2D \operatorname{totvar}[\mathbf{X}]} = \frac{t_{ij}^*}{D \operatorname{totvar}[\mathbf{X}]}. \tag{5.1}$$

Note that with this definition,

$$\sum_{i,j=1}^{D} \tau_{ij} = 1.$$

Then, if $D(D-1)\tau_{ij} > 1$, it means that the logratio of components X_i to X_j contributes to the total variance with a share larger than the average logratio, while a value of $D(D-1)\tau_{ij} < 1$ means that the logratio X_i to X_j has a small contribution to the total variance (smaller than the average); and the smaller τ_{ij} is, the more constant is the logratio X_i to X_j.

5.3   Centering and scaling

In applications, particularly in geology, it is frequent to visualize data in a ternary diagram after rescaling the observations in such a way that their range is roughly the same. This is nothing else but applying a perturbation to the data set, a perturbation that is usually chosen by trial and error. Centering offers an automatic way to improve data representation, without choosing a perturbation. Note that, as mentioned in Proposition 3.3, for a composition, $\mathbf{x}$, and its inverse, $\mathbf{x}^{-1}$, it holds that $\mathbf{x} \oplus \mathbf{x}^{-1} = \mathbf{n}$, the neutral element. This means that perturbation allows to move any composition to the barycenter of the simplex, in the same way that a translation moves real data in real space to the origin. Centering exploits this property. The centering strategy consists of perturbing each row $\mathbf{x}_i = [x_{i1}, x_{i2}, \ldots, x_{iD}]$ of the data set by the inverse of its geometric center $\hat{\mathbf{g}}$

(Definition 5.1). After perturbing each row of the data set by $\hat{g}^{-1}$, the center of the new data set is shifted to the barycenter of the simplex, and the sample will gravitate around this barycenter.

This representation strategy was first introduced by Martín-Fernández et al. (1999) and its potential was shown by Buccianti et al. (1999). An extensive discussion can be found in Eynatten et al. (2002), where it is shown that a perturbation transforms straight lines into straight lines. This allows the inclusion of gridlines and compositional fields in the graphical representation without the risk of a nonlinear distortion. See Figure 5.1 for an example of a data set before and after perturbation with the inverse of the center. The same diagrams show the centering effect on the gridlines.

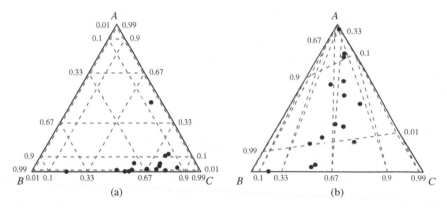

Figure 5.1   Simulated data set (a) before and (b) after centering.

In real space, it is common to standardize a centered variable to unit variance by dividing it by its standard deviation. In a similar way, a (centered) compositional data set can be scaled powering it with totvar$[\mathbf{X}]^{-1/2}$. In this way, a data set with unit total variance is obtained, but preserving the relative contribution of each logratio in the variation array. This is a significant difference with conventional standardization. With real vectors, the relative contributions of each variable are most often artifacts of the units of each variable and, therefore, should be ignored. In contrast, in compositional vectors, all parts share the same *units*, and their relative contribution to the total variation is a rich information that an individual standardization of each part would remove. For this reason, compositional standardization is applied to the data set as a whole, not part by part.

Note that standardization to unit total variance is precisely normalizing the elements of the variation matrix by the average of the total variance discussed in Section 5.2 and shown in Equation (5.1).

5.4   The biplot: a graphical display

Gabriel (1971) introduced the biplot to represent simultaneously the rows and columns of any matrix by means of a rank-2 approximation. Aitchison (1997) adapted it for compositional data and proved it to be a useful exploratory tool. Here we briefly describe the philosophy and mathematics of this technique and, in more depth, its interpretation rules.

5.4.1   Construction of a biplot

Consider the data matrix $\mathbf{X} = [x_{ij}]$ with n rows and D columns, each column corresponding to one part $X_1, X_2, \ldots, X_D$. In other words, D parts have been determined for each one of n samples. Center the data set as described in Section 5.3, and find the clr coefficients $\mathbf{Z}$ (Equation 4.3) of the centered data. Note that $\mathbf{Z}$ is of the same order as $\mathbf{X}$, that is, it has n rows and D columns, and recall that clr coefficients preserve the metrics of the data (Property 4.2) if the vector is considered as a whole. Thus, standard results can be applied to $\mathbf{Z}$ as long as the whole vector is analyzed. In particular, this holds for the spectral decomposition theorem (Eckart and Young, 1936). This implies that the best rank-q approximation $\mathbf{Z}_q^*$ to $\mathbf{Z}$, in the least squares sense, is provided by the singular value decomposition (SVD) of $\mathbf{Z}$ (Krzanowski, 1988, pp. 126–128). Note that this requires the analysis of the whole clr vector, as the equivalence between Aitchison metric and Euclidean metric of the clr vectors does not hold for a subset of clr components.

The SVD of a matrix $\mathbf{Z}$ is obtained from the matrix of eigenvectors $\mathbf{U}$ of $\mathbf{Z}\mathbf{Z}^\mathsf{T}$, from the matrix of eigenvectors $\mathbf{V}$ of $\mathbf{Z}^\mathsf{T}\mathbf{Z}$, and from the square roots of the r, $r \leq \min\{(D-1), n\}$, positive eigenvalues $\lambda_1, \lambda_2, \ldots, \lambda_r$ of either $\mathbf{Z}\mathbf{Z}^\mathsf{T}$ or $\mathbf{Z}^\mathsf{T}\mathbf{Z}$, which are equal up to some null eigenvalues. As a result, taking $k_i = \lambda_i^{1/2}$, it is possible to write

$$\mathbf{Z} = \mathbf{U} \begin{pmatrix} k_1 & 0 & \cdots & 0 \\ 0 & k_2 & \cdots & 0 \\ \vdots & \vdots & \ddots & \vdots \\ 0 & 0 & \cdots & k_r \end{pmatrix} \mathbf{V}^\mathsf{T}, \tag{5.2}$$

where r is the rank of $\mathbf{Z}$, and the singular values $k_1, k_2, \ldots, k_r$ are placed in descending order of magnitude. The matrix $\mathbf{U}$ has dimensions (n, r) and $\mathbf{V}$ is a (D, r)-matrix. Both matrices $\mathbf{U}$ and $\mathbf{V}$ are orthonormal, that is, $\mathbf{U}\mathbf{U}^\mathsf{T} = \mathbf{I}_r$, $\mathbf{V}\mathbf{V}^\mathsf{T} = \mathbf{I}_r$. When $\mathbf{Z}$ is made of centered clr-transformed compositional data, its rows add to zero and, consequently, its rank is $r \leq D - 1$.

The interpretation of the SVD (Equation 5.2) is straightforward. Each row of the matrix $\mathbf{V}^\mathsf{T}$ is the clr of an element of an orthonormal basis of the simplex. In the standard literature of principal component analysis (PCA), these are called *loadings*. This kind of matrices have been denoted by Ψ in Section 4.4. The matrix product $\mathbf{U}\,\mathrm{diag}(k_1, k_2, \ldots, k_r)$ is an (n, r)-matrix, where n rows contain the coordinates of each compositional data point with respect to the orthonormal basis described by $\mathbf{V}^\mathsf{T}$, known as *scores* in PCA. Therefore, $\mathbf{U}\,\mathrm{diag}(k_1, k_2, \ldots, k_r)$ contains ilr-coordinates of the (centered)-compositional data set. Note that these ilr-coordinates are not balances, but general orthonormal coordinates. Singular values $\lambda_1 = k_1^2$, $\lambda_2 = k_2^2$, ..., $\lambda_r = k_r^2$ are proportional to the sample variance of the coordinates.

In order to reduce the dimension of the compositional data set, some orthogonal coordinates can be suppressed, typically those with associated low variance. This can be thought of as a deletion of small square-singular values. Assume the singular values $k_1, k_2, \ldots, k_q$ $(q \le r)$ are retained. Then the proportion of retained variance is

$$\frac{k_1^2 + k_2^2 + \cdots + k_q^2}{k_1^2 + k_2^2 + \cdots + k_r^2} = \frac{\lambda_1 + \lambda_2 + \cdots + \lambda_q}{\lambda_1 + \lambda_2 + \cdots + \lambda_r}.$$

The biplot is normally drawn in two or at most three dimensions, that is, $q = 2$ or $q = 3$ is taken. The resulting biplot will be representative of the data set if the proportion of explained variance is high. The rank-2 approximation is obtained by substituting all singular values with index larger than 2 by zero. The result is

$$\hat{\mathbf{Z}}_2 = \begin{pmatrix} u_{11} & u_{21} \\ u_{12} & u_{22} \\ \vdots & \vdots \\ u_{1n} & u_{2n} \end{pmatrix} \begin{pmatrix} k_1 & 0 \\ 0 & k_2 \end{pmatrix} \begin{pmatrix} v_{11} & v_{21} & \cdots & v_{D1} \\ v_{12} & v_{22} & \cdots & v_{D2} \end{pmatrix}. \tag{5.3}$$

The proportion of variability retained by this approximation is

$$\frac{\lambda_1 + \lambda_2}{\sum_{i=1}^{r} \lambda_i}.$$

To obtain a biplot, it is first necessary to write $\hat{\mathbf{Z}}_2$ as the product of two matrices $\mathbf{AB}^\mathsf{T}$, where $\mathbf{A}$ is an $(n, 2)$ matrix and $\mathbf{B}$ is a $(D, 2)$ matrix. There are different possibilities to obtain such a factorization, depending on the value of $v \in [0, 1]$,

$$\hat{\mathbf{Z}}_2 = \begin{pmatrix} \sqrt{n}k_1^{1-v}\,u_{11} & \sqrt{n}k_2^{1-v}\,u_{21} \\ \sqrt{n}k_1^{1-v}\,u_{12} & \sqrt{n}k_2^{1-v}\,u_{22} \\ \vdots & \vdots \\ \sqrt{n}k_1^{1-v}\,u_{1n} & \sqrt{n}k_2^{1-v}\,u_{2n} \end{pmatrix} \begin{pmatrix} \dfrac{k_1^v v_{11}}{\sqrt{n}} & \dfrac{k_1^v v_{21}}{\sqrt{n}} & \cdots & \dfrac{k_1^v v_{D1}}{\sqrt{n}} \\ \dfrac{k_2^v v_{12}}{\sqrt{n}} & \dfrac{k_2^v v_{22}}{\sqrt{n}} & \cdots & \dfrac{k_2^v v_{D2}}{\sqrt{n}} \end{pmatrix}$$

$$= \begin{pmatrix} \mathbf{a}_1 \\ \mathbf{a}_2 \\ \vdots \\ \mathbf{a}_n \end{pmatrix} \begin{pmatrix} \mathbf{b}_1 & \mathbf{b}_2 & \cdots & \mathbf{b}_D \end{pmatrix}.$$

For $v = 0$, the *covariance biplot* is obtained, where the length of the rays is proportional to the standard deviation of the clr-components, while $v = 1$ leads to the *form biplot*, which is helpful in identifying how good the different clr-components are represented. Ideally, all rays in a form biplot should have length equal to one. Thus, if one ray is much shorter than the others, it is badly represented.

The biplot consists of representing the vectors $\mathbf{a}_i, i = 1, 2 ,..., n$ (n row vectors of two components), and $\mathbf{b}_j, j = 1, 2 ,..., D$ (D column vectors of two components), on a plane. The vectors $\mathbf{a}_1, \mathbf{a}_2, ..., \mathbf{a}_n$ are termed the row markers of $\hat{\mathbf{Z}}_2$ and correspond to the projections of the n samples on the plane defined by the first two eigenvectors of $\mathbf{ZZ}^\top$. The vectors $\mathbf{b}_1, \mathbf{b}_2, ..., \mathbf{b}_D$ are the column markers, which correspond to the projections of the D clr-coefficients on the plane defined by the first two eigenvectors of $\mathbf{Z}^\top\mathbf{Z}$. Both planes are superposed for a visualization of the relationship between data points and variables, in this case clr-coefficients, thus justifying the denomination as *biplot*.

5.4.2   Interpretation of a 2D compositional biplot

The 2D biplot graphically displays the rank-2 approximation $\hat{\mathbf{Z}}_2$ to $\mathbf{Z}$ given by the SVD. A biplot of compositional data consists of the following elements:

- an *origin O* that represents the center of the compositional data set,

- a *vertex* at position $\mathbf{b}_j$ for each of the D clr-variables,

- a *case marker* at position $\mathbf{a}_i$ for each of the n data-points or cases,

- the join, $\overline{O\mathbf{b}_j}$, of O to a vertex $\mathbf{b}_j$, termed a *ray*, and

- the join, $\overline{\mathbf{b}_j\mathbf{b}_k}$, of two vertices $\mathbf{b}_j$ and $\mathbf{b}_k$, termed a *link*.

These features constitute the basic characteristics of a biplot with the following main properties for the interpretation of compositional variability, represented in Figure 5.2 for illustration.

1. The **center** O of the diagram is the centroid (center of gravity) of the D vertices $\mathbf{b}_1, \mathbf{b}_2, ..., \mathbf{b}_D$; as a variable, it represents the geometric mean of the parts used in the clr-transformation; as a datum, it is the center of the data set, $\hat{\mathbf{g}}$, the closed geometric mean of Definition 5.1.

2. **Links** and **rays** provide information on the variability of a logratio in a compositional data set, as

$$|\overline{\mathbf{b}_j\mathbf{b}_k}|^2 \approx \mathrm{var}\left(\ln \frac{X_j}{X_k} \right) \quad \text{and} \quad |\overline{O\mathbf{b}_j}|^2 \approx \mathrm{var}\left(\ln \frac{X_j}{g_m(X)} \right).$$

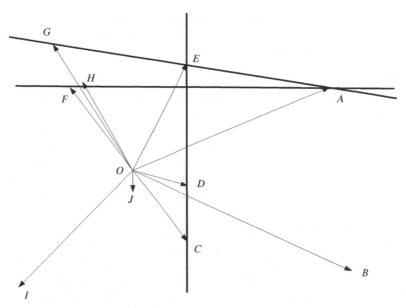

Figure 5.2   Illustration of the interpretation rules of rays and links in a compositional biplot.

Nevertheless, care is needed in interpreting rays, linked to var(clr$_i$(X)). They cannot be identified with var($\ln X_j$), as they depend on the full composition through g_m(**X**), and they also vary when a subcomposition is considered. In other words, interpretation of the rays as indicators of the variance of one single part should be avoided. In Figure 5.2, the clr-coefficient associated to part J has a small variability, while those associated to parts A, B, G, and I have the largest clr-variability. Given that these parts have the largest rays pointing to different directions, they are also going to have long links with most parts of the system, thus large contributions of their logratios to the total variability.

3. **Angles** provide information on the correlation of subcompositions: if links $\overline{\mathbf{b}_j\mathbf{b}_k}$ and $\overline{\mathbf{b}_i\mathbf{b}_\ell}$ intersect at M, then

$$\cos(\overline{\mathbf{b}_j\mathbf{b}_k}M\overline{\mathbf{b}_i\mathbf{b}_\ell}) \approx \mathrm{corr}\left(\ln\frac{X_j}{X_k}, \ln\frac{X_i}{X_\ell}\right).$$

Furthermore, if two links are at right angles, then $\cos(\overline{\mathbf{b}_j\mathbf{b}_k}M\overline{\mathbf{b}_i\mathbf{b}_\ell}) \approx 0$, and zero correlation of the two logratios can be expected. An example in Figure 5.2 is given by the subcompositions $\{A, F, H\}$ and $\{C, D, E\}$, implying that any balance between groups of parts of $\{A, F, H\}$ and any balance between groups of parts of $\{C, D, E\}$ should be checked for possible zero correlation.

4. **Subcompositional analysis.** Ratios are preserved under formation of subcompositions; it follows that the biplot for any subcomposition is simply formed by selecting the vertices corresponding to the parts of the subcomposition and taking the center of the subcompositional biplot as the centroid of these vertices. Nevertheless, one must be aware that the biplot is a projection, and it might not be the same as the one obtained when computing the biplot for the chosen subcomposition. It depends on how good the subcomposition is represented in the global biplot.

5. **Coincident vertices.** If vertices $\mathbf{b}_j$ and $\mathbf{b}_k$ coincide, or nearly so, this means that $\mathrm{var}(\ln(X_j/X_k))$ is zero, or nearly so, and the ratio X_j/X_k is constant, or nearly so. Then, the two involved parts, X_j and X_k, can be assumed to be redundant. If the proportion of variance captured by the biplot is not very high, two coincident vertices suggest that $\ln(X_j/X_k)$ is orthogonal to the plane of the biplot, and this might be an indication of the possible independence of that logratio with the first two principal directions of the SVD. This is the case of vertices H and F in Figure 5.2. For a test on proportionality of two parts, see Egozcue et al. (2013).

6. **Collinear vertices.** If a subset of vertices is collinear, this might indicate that the associated subcomposition has a biplot that is one dimensional, which might mean that the subcomposition has one-dimensional variability, that is, compositions plot along a compositional line. Figure 5.2 shows three examples of these associations: $\{A, E, G\}$, $\{C, D, E\}$, and $\{A, F, H\}$, although the last is not perfectly collinear.

Markers, placed at $\mathbf{a}_1, \mathbf{a}_2, \ldots, \mathbf{a}_n$ (see Figure 5.3), are related to each individual observation. They can eventually be colored by groups (or different symbols). Also, group averages or further data not used for the biplot calculations can be represented. In both cases, the whole composition of that particular observation/average can be approximated from the relative position of its marker with respect to the ray constellation. Every ray (or link) of the biplot defines an axis of representation of the related clr-component (or pairwise logratio), increasing in the direction of the arrow and decreasing in the opposite sense (links are not physically represented as arrows by default, but the same ideas apply to them, with the arrow pointing at the part in the numerator). Given the rules relating to the center and the length of arrows mentioned before, each marker can be interpreted by orthogonally projecting it onto each axis, with the following rules, illustrated in

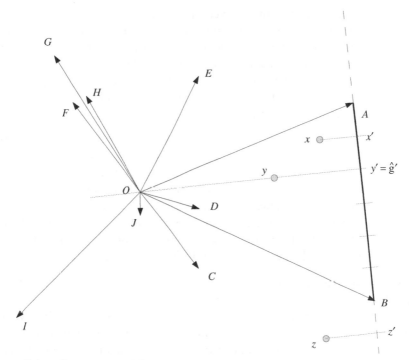

Figure 5.3   Illustration of the interpretation rules of markers in a compositional biplot.

Figure 5.3. In this figure, three markers are represented as dots x, y, and z, and the center of the data set $\hat{g}$ is placed at the center of the diagram O. For illustration, the link is represented as $\overline{AB}$, as well as the projected markers as x', y', z', and $\hat{g}'$.

1. If the projected marker falls on the *tail* of the arrow of the clr-component i (near the center O of the diagram), that particular observation has an ith clr component close or equal to the average of the data set.

2. When dealing with a link between parts i and j, if the projections of the marker and the center fall onto the same place on that axis, then that particular observation has a value of $\ln(X_i/X_j)$ equal to its average on the data set; this is the case of sample y projected onto the axis AB, meaning that the value of $\ln(A/B)$ for that sample should coincide (or nearly so) with the average of that logratio in the whole data set.

3. If the projected marker falls at the *end* of the ray or arrow of variable i, that particular observation has an ith clr component approximately one standard deviation larger than the average.

4. When dealing with a link between parts i and j, if the projections of the marker and the center fall one link-length apart from each other, then that particular observation has a value on $\ln(X_i/X_j)$ one standard deviation larger than the average of the data set; this is the case of sample x, where projection x' onto the AB axis falls roughly at $(1/6)\overline{AB}$ off the projection $\hat{g}'$ of $\hat{g}$: this means that the value of the logratio $\ln(A/B)$ for sample x is one-sixth of the standard deviation larger than the average value.

5. In the same way, markers projected on the opposite direction of an arrow have clr values or logratios smaller than the average; the length of rays and links gives approximately the equivalent of one standard deviation to measure how much more/less that particular observation has in a given logratio with respect to the average of the data set. This case is represented by sample z, which position suggests that this should be $\approx 5/6$ of the standard deviation smaller in $\ln(A/B)$ than the average or, inversely, it should be $\approx 5/6$ standard deviations larger in $\ln(B/A)$ than the average.

Interpreting individual markers seldom brings much insight into the data set nature, but sometimes it does. In the presence of a few outliers, the biplot might be dominated by these values, and ratios can be found that actually behave in an anomalous way. That will guide a preliminary outlier analysis and suggest the removal of the outliers themselves *or* of the components in which they behave abnormally. Group means can also be plotted. They offer an idea of the main ratios that distinguish them: one should nevertheless always remember that an SVD biplot is not a discriminating tool, as it is not optimized for this group separation.

These rules are used in some worked examples in Section 5.6. From these aspects of interpretation, it should be clear that links are fundamental elements of a compositional biplot. The lengths of links are (approximately) proportional to the standard deviations of pairwise logratios, as they appear (squared) in the variation matrix. The constellation of links informs about the compositional covariance structure of the composition and provides hints about subcompositional variability and independence. Note that the interpretation rules of a biplot use only the shape of the diagram, and this is unaffected by any rotation or mirror imaging of the diagram.

For some applications of biplots to compositional data in a variety of geological contexts, see Aitchison (1990), and for a deeper insight into biplots of compositional data, with applications in other disciplines and extensions to conditional biplots, see Aitchison and Greenacre (2002).

5.5   Exploratory analysis of coordinates

Either as a result of the preceding descriptive analysis or because of a prior knowledge of the problem at hand, a given sequential binary partition (see

Section 4.5) can be considered as particularly interesting. In this case, its associated orthonormal coordinates, being a vector of real variables, can be treated with the existing battery of conventional descriptive analysis techniques. If $\mathbf{X}^* = \mathrm{ilr}(\mathbf{X})$ denotes the coordinates of the data set – rows contain the coordinates of observations – then its experimental moments satisfy

$$\bar{\mathbf{x}}^* = \mathrm{ilr}(\hat{\mathbf{g}}) = \mathrm{clr}(\hat{\mathbf{g}}) \times \boldsymbol{\varPsi}^\top = \ln(\hat{\mathbf{g}}) \times \boldsymbol{\varPsi}^\top, \tag{5.4}$$

$$\mathbf{S} = -\boldsymbol{\varPsi} \times \mathbf{T}^* \times \boldsymbol{\varPsi}^\top = -\frac{1}{2}\boldsymbol{\varPsi} \times \mathbf{T} \times \boldsymbol{\varPsi}^\top, \tag{5.5}$$

with $\boldsymbol{\varPsi}$ the contrast matrix (see Section 4.5 for its construction); $\hat{\mathbf{g}}$ the center of the data set as given in Definition 5.1; and $\mathbf{T}^*$ or $\mathbf{T}$ the (normalized) variation matrix as introduced in Definition 5.2. Correlation matrices $\mathbf{R} = (r_{ij})$ can be computed from $\mathbf{S} = (s_{ij})$ with

$$r_{ij} = \frac{s_{ij}}{\sqrt{s_i s_j}}$$

and can be interpreted with standard rules of linear association: r_{ij} close to 0 means lack of linear association between the coordinates X_i^* and X_j^*, while the closer r_{ij} to ± 1, the stronger the linear association is.

A graphical representation, with the specific aim of representing a system of coordinates based on a sequential binary partition, is the CoDa- or balance-dendrogram (Egozcue and Pawlowsky-Glahn, 2006; Pawlowsky-Glahn and Egozcue, 2006, 2011). A *balance-dendrogram* is the joint representation of the following elements:

1. a sequential binary partition, in the form of a tree structure;

2. the sample variance and mean of each balance, as vertical bars and inception points of these bars;

3. a box-plot, summarizing the order statistics of each balance on horizontal axes.

Each coordinate is represented in the horizontal axis, which limits correspond to a certain range (fixed by the user, the same for each coordinate). The vertical bar going up from each one of these coordinate axes represents the variance of that specific coordinate, and the contact point is the coordinate mean. Figure 5.4 shows these elements in an illustrative example.

Given that the range of each coordinate is symmetric (in Figure 5.4, it goes from -3 to $+3$), the box plots closer to one part (or group) indicate that that part (or group) is more abundant. Thus, in Figure 5.4, SiO_2 is slightly more abundant than Al_2O_3, there is more FeO than Fe_2O_3, and much more structural oxides (SiO_2 and Al_2O_3) than the rest. Another feature easily read from a balance-dendrogram is symmetry: it can be assessed both by comparison between the several quantile boxes, and by looking at the difference between the median (marked as Q2 in Figure 5.4b) and the mean.

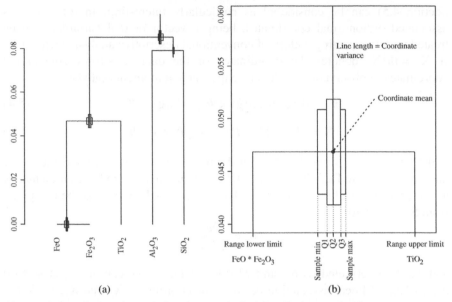

*Figure 5.4   Illustration of elements included in a balance-dendrogram.
(a) represents a full dendrogram, and (b) is an expanded symbolic
representation of the balance of* (FeO,Fe$_2$O$_3$) *against* TiO$_2$.

As mentioned, coordinates can be analyzed with the standard set of tools for
real-valued variables. In particular, biplots can be derived from their covariance
structure. Using the SVD explained in Section 5.4.1 and replacing $\mathbf{Z}$ by the matrix
of coordinates $\mathbf{X}^*$, one can derive a rank-2 best approximation $\hat{\mathbf{X}}_2^*$, which allows a
decomposition $\hat{\mathbf{X}}_2^* = \mathbf{A}\,\mathbf{B}^\mathsf{T}$ in markers and rays. The markers $\mathbf{A}$ happen to be the
same as before, and the rays show the relations between the chosen coordinates,
with the following rules:

1. the **length** of a ray provides information about the variance of that coordinate: the longer the ray, the larger the variance,

$$|\overline{Ob_i}|^2 \approx \mathrm{var}\left(\ln\mathbf{X}\,\boldsymbol{\psi}_i^\mathsf{T}\right);$$

2. the **angle** between two rays informs of the correlation between coordinates,

$$\cos(\mathbf{b}_i O\mathbf{b}_j) \approx \mathrm{corr}\left(\ln\mathbf{X}\,\boldsymbol{\psi}_i^\mathsf{T}, \ln\mathbf{X}\,\boldsymbol{\psi}_j^\mathsf{T}\right),$$

which implies that orthogonal links suggest uncorrelated coordinates and
nearly parallel links suggest well-correlated coordinates;

3. opposite to a compositional biplot, the links of a coordinate biplot cannot be interpreted in general terms: the link $\overline{\mathbf{b}_i\mathbf{b}_j}$ represents the difference between the two coordinates, that is, $\ln \mathbf{X}\,(\boldsymbol{\psi}_i - \boldsymbol{\psi}_j)^{\mathsf{T}}$; if $(\boldsymbol{\psi}_i - \boldsymbol{\psi}_j)$ has an interpretation, then it will be worth understanding the length of the link as indication of the standard deviation of this difference, and its angle with another ray or link as an indication of uncorrelation/correlation.

Markers can be interpreted with the same rules given in Section 5.4.2 (in particular Figure 5.3) by projecting them onto the rays representing each coordinate.

5.6   A geological example

The first illustrative example consists of 17 samples of basaltic rocks from Kilauea Iki lava lake, Hawaii (Table 5.1). This data set $\mathbf{X}$, first published by Richter and Moore (1966), is reproduced in Rollinson (1995). Originally, 14 parts were registered, but H_2O^+ and H_2O^- have been omitted because of their

Table 5.1   Chemical analysis of rocks from Kilauea Iki lava lake, Hawaii.

SiO_2	TiO_2	Al_2O_3	Fe_2O_3	FeO	MnO	MgO	CaO	Na_2O	K_2O	P_2O_5	CO_2
48.29	2.33	11.48	1.59	10.03	0.18	13.58	9.85	1.90	0.44	0.23	0.01
48.83	2.47	12.38	2.15	9.41	0.17	11.08	10.64	2.02	0.47	0.24	0.00
45.61	1.70	8.33	2.12	10.02	0.17	23.06	6.98	1.33	0.32	0.16	0.00
45.50	1.54	8.17	1.60	10.44	0.17	23.87	6.79	1.28	0.31	0.15	0.00
49.27	3.30	12.10	1.77	9.89	0.17	10.46	9.65	2.25	0.65	0.30	0.00
46.53	1.99	9.49	2.16	9.79	0.18	19.28	8.18	1.54	0.38	0.18	0.11
48.12	2.34	11.43	2.26	9.46	0.18	13.65	9.87	1.89	0.46	0.22	0.04
47.93	2.32	11.18	2.46	9.36	0.18	14.33	9.64	1.86	0.45	0.21	0.02
46.96	2.01	9.90	2.13	9.72	0.18	18.31	8.58	1.58	0.37	0.19	0.00
49.16	2.73	12.54	1.83	10.02	0.18	10.05	10.55	2.09	0.56	0.26	0.00
48.41	2.47	11.80	2.81	8.91	0.18	12.52	10.18	1.93	0.48	0.23	0.00
47.90	2.24	11.17	2.41	9.36	0.18	14.64	9.58	1.82	0.41	0.21	0.01
48.45	2.35	11.64	1.04	10.37	0.18	13.23	10.13	1.89	0.45	0.23	0.00
48.98	2.48	12.05	1.39	10.17	0.18	11.18	10.83	1.73	0.80	0.24	0.01
48.74	2.44	11.60	1.38	10.18	0.18	12.35	10.45	1.67	0.79	0.23	0.01
49.61	3.03	12.91	1.60	9.68	0.17	8.84	10.96	2.24	0.55	0.27	0.01
49.20	2.50	12.32	1.26	10.13	0.18	10.51	11.05	2.02	0.48	0.23	0.01

large amount of zeros. CO_2 has been kept in the table, to call attention upon this zero issue, but has been omitted from the study. This is the strategy to follow if the part with zeros is not essential in the characterization of the phenomenon under study. If the part is essential and the proportion of zeros is small, then imputation techniques should be used. They replace the missing value by a suitable (usually small) value, as explained in Section 2.3.2. The case where the part is essential and the proportion of zeros high is treated in the example of Section 5.9.

The center of this data set (excluding CO_2) is

$$\hat{g} = (48.57, 2.35, 11.23, 1.84, 9.91, 0.18, 13.74, 9.65, 1.82, 0.48, 0.22),$$

the total variance is $\text{totvar}[\mathbf{X}] = 0.3275$, and the normalized variation matrix $\mathbf{T}^*$ is given in Table 5.2.

The biplot in Figure 5.5 shows a two-dimensional pattern of variability: two sets of parts that cluster together, $A = [TiO_2, Al_2O_3, CaO, Na_2O, P_2O_5]$ and $B = [SiO_2, FeO, MnO]$, and a one-dimensional relationships between them.

The two-dimensional pattern of variability is supported by the fact that the first two axes of the biplot reproduce about 90% of the total variance, as captured in the scree plot in Figure 5.5a. The orthogonality of the link between Fe_2O_3 and FeO (i.e., the oxidation state) with the link between MgO and any of the parts in set A might help in finding an explanation for this behavior and in decomposing the global pattern into two independent processes.

Concerning the two sets of parts, we can observe short links between them and, at the same time, the variances of the corresponding logratios (see the normalized

Table 5.2   Normalized variation matrix of data in Table 5.1. For simplicity, only the upper triangle is represented, omitting the first column and last row.

t_{ij}^*	TiO_2	Al_2O_3	Fe_2O_3	FeO	MnO	MgO	CaO	Na_2O	K_2O	P_2O_5
SiO_2	0.012	0.006	0.036	0.001	0.001	0.046	0.007	0.009	0.029	0.011
TiO_2		0.003	0.058	0.019	0.016	0.103	0.005	0.002	0.015	0.000
Al_2O_3			0.050	0.011	0.008	0.084	0.000	0.002	0.017	0.002
Fe_2O_3				0.044	0.035	0.053	0.054	0.050	0.093	0.059
FeO					0.001	0.038	0.012	0.015	0.034	0.017
MnO						0.040	0.009	0.012	0.033	0.015
MgO							0.086	0.092	0.130	0.100
CaO								0.003	0.016	0.004
Na_2O									0.024	0.002
K_2O										0.014

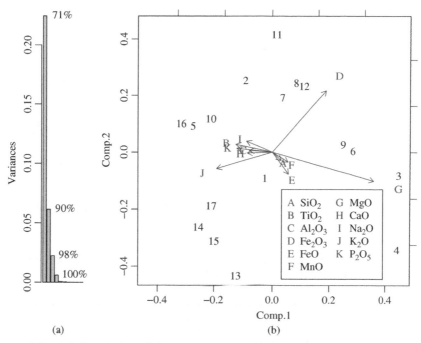

Figure 5.5   (a) Scree plot of the variances of all principal components, with indication of explained cumulative variance, and (b) biplot of data of Table 5.1. Variables are clr *transformed.*

variation matrix $\mathbf{T}^*$, Table 5.2) are very close to zero. Consequently, we can say that they are mostly redundant and that some of them could be either grouped in a single part or simply omitted. In both cases, the dimensionality of the problem would be reduced.

Another aspect to be considered is the diverse patterns of one-dimensional variability that can be observed. Examples of three-part compositions, which can be visualized in a ternary diagram, are Fe_2O_3, K_2O, and any of the parts in set A, or MgO with any of the parts in set A and any of the parts in set B. Take, for example, Fe_2O_3, K_2O, and Na_2O. After closure, the samples plot in a ternary diagram as shown in Figure 5.6 and the expected trend can be recognized, as well as two outliers corresponding to samples 14 and 15, which would require further analysis. Regarding the trend itself, note that it is close to a line of isoproportion Na_2O/K_2O: it can be concluded that the ratio of these two parts is independent of the amount of Fe_2O_3.

As a last step, conventional descriptive statistics of the orthonormal coordinates in a specific reference system (either chosen or derived from the previous steps) are computed. In this case, knowledge of the typical geochemistry and

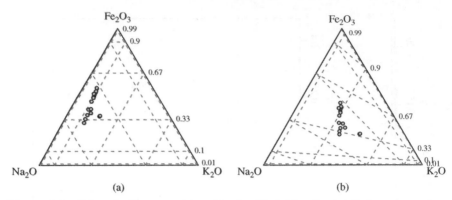

Figure 5.6   Plot of subcomposition (Fe_2O_3, K_2O, Na_2O). *(a) Before centering.*
(b) After centering.

Table 5.3   A possible sequential binary partition for the data set of Table 5.1.
Note that some packages, such as CoDaPack, use a reverse order to define such
a partition.

	SiO_2	TiO_2	Al_2O_3	Fe_2O_3	FeO	MnO	MgO	CaO	Na_2O	K_2O	P_2O_5
b_1	0	0	0	+1	−1	0	0	0	0	0	0
b_2	+1	0	−1	0	0	0	0	0	0	0	0
b_3	0	+1	0	0	0	0	0	0	0	0	−1
b_4	+1	−1	+1	0	0	0	0	0	0	0	−1
b_5	0	0	0	0	0	0	0	+1	−1	0	0
b_6	0	0	0	0	0	0	0	+1	+1	−1	0
b_7	0	0	0	0	0	+1	−1	0	0	0	0
b_8	0	0	0	+1	+1	−1	−1	0	0	0	0
b_9	0	0	0	+1	+1	+1	+1	−1	−1	−1	0
b_{10}	+1	+1	+1	−1	−1	−1	−1	−1	−1	−1	+1

mineralogy of basaltic rocks suggests a priori the set of balances in Table 5.3,
where the resulting balances can be interpreted as

1. an oxidation state proxy (Fe^{3+} against Fe^{2+});

2. a silica saturation proxy (when SiO_2 is lacking, Al_2O_3 takes its place);

3. distribution within heavy minerals (rutile TiO_2 rich or apatite P_2O_5 rich?);

4. importance of heavy minerals relative to silicates;

5. distribution within plagioclase (albite or anorthite?);

6. distribution within feldspar (K_2O-feldspar or CaO/Na_2O-rich plagioclase?);

7. distribution within mafic nonferric minerals;

8. distribution within mafic minerals (ferric vs. nonferric);

9. importance of mafic minerals against feldspar;

10. importance of cation oxides (those filling the crystalline structure of minerals) against frame oxides (mainly Al_2O_3 and SiO_2).

One should be aware that such an interpretation is totally problem driven; for example, working with sedimentary rocks, it would have no sense to split MgO and CaO (as they would mainly occur in limestones and associated lithologies) or to group Na_2O with CaO (as they would probably come from different rock types, e.g., siliciclastic against carbonate).

The sequential binary partition of Table 5.3 is represented as a balance-dendrogram of the sample (Figure 5.7) with the range $(-3, +3)$. This range translates the two-part compositions to proportions of $(1.4, 98.6)\%$; for example, looking at the balance MgO–MnO, the variance bar is placed at the lower extreme of the balance axis, which implies that in this subcomposition, MgO represents in average more than 98% and MnO less than 2%. Looking at the lengths of the several variance bars, it can be observed that the balances P_2O_5–TiO_2 and SiO_2–Al_2O_3 are almost constant, as their bars are very short and their box plots extremely narrow. The distribution between K_2O, Na_2O, and

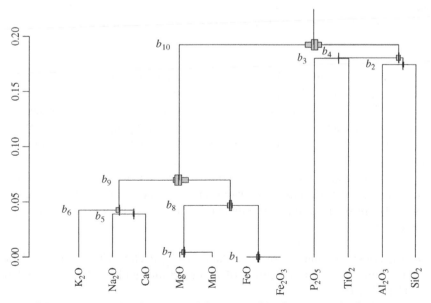

Figure 5.7   Balance-dendrogram of data in Table 5.1 using the balances of Table 5.3.

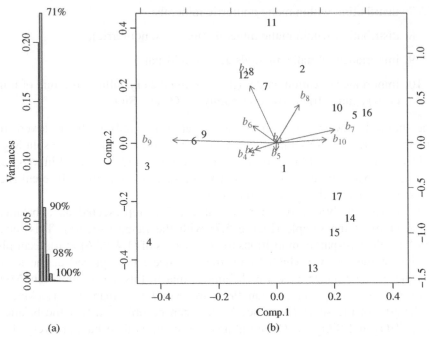

Figure 5.8 *(a) Scree plot of the variances of all principal components, with indication of cumulative explained variance, and (b) biplot of data of Table 5.1 expressed in the balance coordinate system of Table 5.3. Compare with Figure 5.5, in particular the scree plot, the configuration of data points, and the links between the parts related to balances b_1, b_2, b_3, b_5, and b_7.*

CaO tells that Na_2O and CaO keep a quite constant ratio (consequently, there are no strong variations in the plagioclase composition), and the ratio of these two against K_2O is also fairly constant, with the exception of some values below the first quartile (probably, a single value with a particularly high K_2O content). The other balances are well equilibrated (in particular, see how centered is the proxy balance between feldspar and mafic minerals), all with moderate dispersion.

Once the marginal empirical distribution of the balances has been analyzed, the biplot (Figure 5.8), and the conventional covariance or correlation matrices (Table 5.4) can be used to explore their relations. From these, the following can be observed:

- The constant behavior of b_3 (balance TiO_2/P_2O_5), with a variance below 10^{-4} and, in a lesser degree, of b_5 (anorthite–albite relation, or balance CaO/Na_2O).

Table 5.4   Covariance (lower triangle) and correlation (upper triangle) array of balances

	b_1	b_2	b_3	b_4	b_5	b_6	b_7	b_8	b_9	b_{10}
b_1	0.047	0.120	0.341	0.111	−0.283	0.358	−0.212	0.557	0.423	−0.387
b_2	0.002	0.006	−0.125	0.788	0.077	0.234	−0.979	−0.695	0.920	−0.899
b_3	0.002	−0.000	0.000	−0.345	−0.380	0.018	0.181	0.423	−0.091	0.141
b_4	0.003	0.007	−0.001	0.012	0.461	0.365	−0.832	−0.663	0.821	−0.882
b_5	−0.004	0.000	−0.000	0.003	0.003	−0.450	−0.087	−0.385	−0.029	−0.275
b_6	0.013	0.003	0.000	0.007	−0.004	0.027	−0.328	−0.029	0.505	−0.243
b_7	−0.009	−0.016	0.001	−0.019	−0.001	−0.011	0.042	0.668	−0.961	0.936
b_8	0.018	−0.008	0.001	−0.011	−0.003	−0.001	0.021	0.023	−0.483	0.516
b_9	0.032	0.025	−0.001	0.031	−0.001	0.029	−0.069	−0.026	0.123	−0.936
b_{10}	−0.015	−0.013	0.001	−0.017	−0.003	−0.007	0.035	0.014	−0.059	0.032

- Orthogonality of the pairs of rays b_1–b_2, b_1–b_4, b_1–b_7, and b_6–b_8 suggests the lack of correlation of their balances, confirmed by Table 5.4, where correlations of less than ± 0.3 are reported. In particular, the pair b_6–b_8 has a correlation of −0.029. These facts would imply that silica saturation (b_2), the presence of heavy minerals (b_4), and the MnO/MgO balance (b_7) are uncorrelated with the oxidation state (b_1) and that the type of feldspars (b_6) is unrelated to the type of mafic minerals (b_8).

- Balances b_9 and b_{10} are opposite, and their correlation is −0.936, implying that the ratio mafic oxides/feldspar oxides is high when the ratio silica–alumina/cation oxides is low, that is, mafics are poorer in silica and alumina.

A final comment regarding balance descriptive statistics: because the balances are chosen according to their interpretability, we are no more just *describing* patterns here. Balance statistics represent a step further toward modeling: all conclusions in these last three points heavily depend on the preliminary interpretation (= *model*) of the computed balances.

5.7   Linear trends along principal components

The SVD applied to centered clr-data has been presented in Section 5.4 as a technique for dimension reduction of compositional data. As a result, the compositional biplot has been shown to be a powerful exploratory tool. In a statistical framework, this technique is known as principal component analysis (PCA). Additionally, PCA–SVD can be used as the appropriate modeling tool whenever the presence of a trend in the compositional data set is suspected, but no external controlling variable has been reported (e.g., Eynatten et al., 2003; Tolosana-Delgado et al., 2005). In this case, one can express the SVD (Equation 5.2) using operations of the simplex, where each row of the data set becomes

$$\mathbf{x}_n = \hat{\mathbf{g}} \oplus \bigoplus_{i=1}^{r}(u_{ni}k_i) \odot \mathrm{clr}^{-1}(\mathbf{v}_i),$$

with $\mathbf{v}_i$ representing a row of matrix $\mathbf{V}$. If the proportion of variability explained by the first principal component (PC) is large (say $> 95\%$), then one can approximate each composition of the data set by a rank-1 approximation (Eckart and Young, 1936),

$$\hat{\mathbf{x}}_n = \hat{\mathbf{g}} \oplus (u_{n1}k_1) \odot \mathrm{clr}^{-1}(\mathbf{v}_1) = \hat{\mathbf{g}} \oplus t_n \odot \mathbf{w}_1, \qquad (5.6)$$

being $t_n = u_{n1}k_1$ the latent trend and denoting $\mathbf{w}_1 = \mathrm{clr}^{-1}(\mathbf{v}_1)$ for simplicity. In the case that external covariables are actually available, one should rather apply regression techniques, explained in Section 8.1.

Example 5.4. To illustrate this technique, let us consider the simplest case, in which one PC (i.e., one square-singular value) explains a large proportion of the total variance, in this case more than 98%, such as the one in Figure 5.6. It corresponds to the subcomposition $[Fe_2O_3,K_2O,Na_2O]$ from Table 5.1, after removing samples 14 and 15. The center of this set is $\hat{\mathbf{g}} = [0.455, 0.107, 0.438]$. In Figure 5.9, the compositional line going through the barycenter of the simplex along $\mathbf{w}_1 = [0.63, 18, 19]$, describes the trend shown by the centered sample,

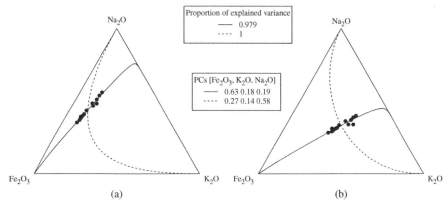

Figure 5.9   Principal components in S^3. (a) Before centering. (b) After centering.

and $\hat{\mathbf{g}} \oplus t \odot \mathbf{w}_1$ describes the trend shown in the noncentered data set. The figure also contains the second compositional PC, $\mathbf{w}_2 = \text{clr}^{-1}(\mathbf{v}_2) = [0.27, 0.14, 0.58]$, which carries only 2% of the variability within this subcomposition. Thus, it can be considered to be constant, that is,

$$\langle \mathbf{w}_2, \mathbf{X} \rangle_a = C.$$

Given that

$$\text{clr}(\mathbf{w}_2) = [-0.036, -0.688, 0.724] \approx [0, -0.707, 0.707] = 2^{-1/2}[0, -1, +1],$$

we can write

$$-\frac{1}{\sqrt{2}} \ln K_2O + \frac{1}{\sqrt{2}} \ln Na_2O \approx \hat{C} = \langle \mathbf{w}_2, \hat{\mathbf{g}} \rangle_a,$$

with $\hat{C} = 0.970$ if calculated with the true $\mathbf{w}_2$ and $\hat{C} = 0.998$ using the approximate $\mathbf{w}_2 \approx 2^{-1/2}[0, -1, +1]$. This is equivalent to

$$\ln \frac{Na_2O}{K_2O} = \sqrt{2}\hat{C} = 1.411, \qquad \frac{Na_2O}{K_2O} = e^{1.411} = 4.100,$$

that is, the ratio of Na_2O to K_2O is roughly constant and approximately 4. The evolution of the proportion per unit volume of each part, as described by the first PC, is shown in Figure 5.10a. The cumulative proportion is drawn in Figure 5.10b. These diagrams were obtained by using (Equation 5.6) with a sequence of t values from -3 to $+3$.   ◇

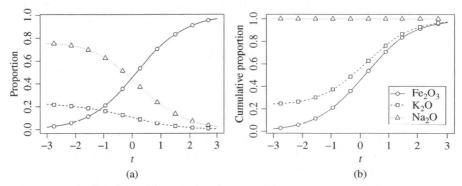

Figure 5.10 Evolution of proportions as described by the first principal component. (a) Proportions. (b) Cumulative proportions.

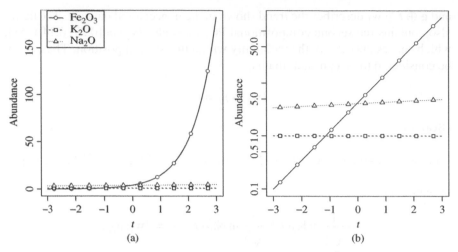

Figure 5.11   Interpretation of a principal component in S^3 under the assumption of stability of K_2O; the vertical scale is represented in (a) raw scale and (b) log scale.

To interpret a trend, we can use Equation (3.1), which allows us to rescale the direction of the first PC, assuming whatever is convenient according to the process under study, for example, that one part is stable. A representation of the result is also possible, as can be seen in Figure 5.11. The part assumed to be stable, K_2O, has a constant, unit perturbation coefficient. We see that under this assumption, within the range of variation of the observations, Na_2O has only a very small increase, while Fe_2O_3 shows a considerable increase compared to the other two parts. In other words, one possible explanation for the observed pattern of variability is that Fe_2O_3 varies significantly, while the other two parts remain stable. All these increases are linear in log scale. The graph gives additional information: the relative behavior will be preserved under any assumption. If the assumption is that K_2O increases (decreases), then Na_2O will roughly show the same behavior as K_2O, while Fe_2O_3 will always change from *below* to *above* the reference, constant level.

Note that although a ternary diagram can represent a perturbation process described by a PC only in three components, we can extend the representation used in Figure 5.10 to as many parts as we are interested in (for an example with five parts, see Figure 5.12). As the number of parts represented increases, it becomes more difficult that the first PC represents such a high proportion of variance as suggested before. In this case, it is sensible to represent the data as well, for the sake of comparison.

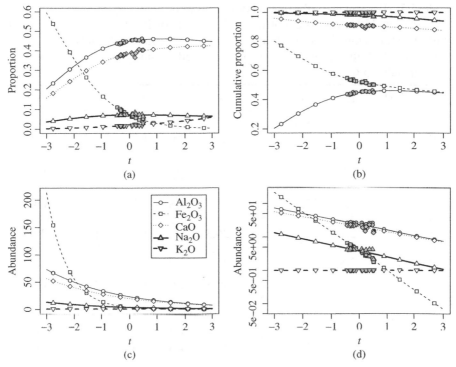

Figure 5.12   Interpretation of a principal component in S^5; the upper row shows the evolution of (a) proportions and (b) cumulative proportions; and the lower row the evolution of abundances under the assumption of stability of K_2O with the vertical scale represented in (c) raw scale and (d) log scale. Gray-shaded symbols correspond to actual data; curves follow the rank-1 approximated trend. This one-dimensional trend represents 82% of the total variability of the five parts.

5.8   A nutrition example

In Pawlowsky-Glahn and Egozcue (2011), a data set of consumption of several food categories in Europe in the early 1980s is presented (reproduced from Peña (1991)). The data were originally reported in tons per year of consumed food in several categories: red meats (pork, beef, veal), white meats (chicken, turkey), eggs, milk products, fish (sea and freshwater), cereals, starch sources (potatoes, rice), nuts, and vegetables (including fruit and orchard products). Twenty five countries are considered, whose names and abbreviations are reported in Table 5.5, together with the percentages of protein consumption provided by each of the nine food categories. Note that the fill up variable to 100% is not reported.

Table 5.5   Percentages of protein consumption provided by each of nine food categories in 25 European countries in the decade of 1980. RM, red meat; WM, white meat; E, eggs; M, milk; F, Fish; C, cereals; S, starch; N, nuts; V, vegetables. Fill up variable to 100% not reported. Country names and groups given in Table 5.6.

RM	WM	E	M	F	C	S	N	V	Abbreviation
10.1	1.4	0.5	8.9	0.2	42.3	0.6	5.5	1.7	Al
8.9	14.0	4.3	19.9	2.1	28.0	3.6	1.3	4.3	Au
13.5	9.3	4.1	17.5	4.5	26.6	5.7	2.1	4.0	Be
7.8	6.0	1.6	8.3	1.2	56.7	1.1	3.7	4.2	Bu
9.7	11.4	2.8	12.5	2.0	34.3	5.0	1.1	4.0	Cz
10.6	10.8	3.7	25.0	9.9	21.9	4.8	0.7	2.4	Dk
8.4	11.6	3.7	11.1	5.4	24.6	6.5	0.8	3.6	Gw
9.5	4.9	2.7	33.7	5.8	26.3	5.1	1.0	1.4	Fi
18.0	9.9	3.3	19.5	5.7	28.1	4.8	2.4	6.5	Fr
10.2	3.0	2.8	17.6	5.9	41.7	2.2	7.8	6.5	Gr
5.3	12.4	2.9	9.7	0.3	40.1	4.0	5.4	4.2	Hn
13.9	10.0	4.7	25.8	2.2	24.0	6.2	1.6	2.9	Ei
9.0	5.1	2.9	13.7	3.4	36.8	2.1	4.3	6.7	It
9.5	13.6	3.6	23.4	2.5	22.4	4.2	1.8	3.7	Nt
9.4	4.7	2.7	23.3	9.7	23.0	4.6	1.6	2.7	Nw
6.9	10.2	2.7	19.3	3.0	36.1	5.9	2.0	6.6	Pl
6.2	3.7	1.1	4.9	14.2	27.0	5.9	4.7	7.9	Po
6.2	6.3	1.5	11.1	1.0	49.6	3.1	5.3	2.8	Ro
7.1	3.4	3.1	8.6	7.0	29.2	5.7	5.9	7.2	Es
9.9	7.8	3.5	24.7	7.5	19.5	3.7	1.4	2.0	Sw
13.1	10.1	3.1	23.8	2.3	25.6	2.8	2.4	4.9	Sz
17.4	5.7	4.7	20.6	4.3	24.3	4.7	3.4	3.3	UK
9.3	4.6	2.1	16.6	3.0	43.6	6.4	3.4	2.9	USR
11.4	12.5	4.1	18.8	3.4	18.6	5.2	1.5	3.8	Ge
4.4	5.0	1.2	9.5	0.6	55.9	3.0	5.7	3.2	Yu

   The goal of this analysis is to identify nutrition patterns throughout Europe and, eventually, group the reported countries in common nutrition patterns. The data is quite old, from the time when the communist block and the Iron Curtain still existed, but this analysis will be illustrative of a CoDa exploratory methodology, even if conclusions are a bit obsolete.

   The first steps provide mean (center) and variability estimates reported in Table 5.7. Not surprisingly, in the subcomposition of nine parts, proteins

Table 5.6   Country names and groups of the
abbreviations given in Table 5.5. EW=1 for Eastern
countries, EW=2 for Western countries; NS=1 for
Northern countries, NS=2 for Southern countries.

Land	Abbreviation	EW	NS
Albania	Al	1	2
Austria	Au	2	1
Belgium	Be	2	1
Bulgaria	Bu	1	2
Czech Republic	Cz	1	1
Denmark	Dk	2	1
Western Germany	Gw	1	1
Finland	Fi	2	1
France	Fr	2	2
Greece	Gr	2	2
Hungary	Hn	1	1
Ireland	Ei	2	1
Italy	It	2	2
Netherlands	Nt	2	1
Norway	Nw	2	1
Poland	Pl	1	1
Portugal	Po	2	2
Romania	Ro	1	2
Spain	Es	2	2
Sweden	Sw	2	1
Switzerland	Sz	2	1
United Kingdom	UK	2	1
USSR	USR	1	1
Eastern Germany	Ge	2	1
Yugoslavia	Yu	1	2

provided by cereals (∼40%) and milk products (∼20%) dominate the average diet in Europe, while protein sources account for a global percentage around 25%, with more red than white meat, and less eggs or fish (with a similar 3% contribution). Vegetables, starch, and nuts have a similar average consumption.

With regard to variability, the smallest contributions are represented by the red meats to milk logratio (0.14) and the white meat to eggs logratio (0.18), thus these

Table 5.7   Descriptive statistics of the nutrition data of Table 5.5

Statistic	RM	WM	E	M	F	C	S	N	V
Mean (center)	11.93	8.85	3.4	19.94	3.71	39.28	4.89	3.17	4.82
Variation	RM	WM	E	M	F	C	S	N	V
RM		0.38	0.22	0.14	0.98	0.33	0.40	0.76	0.35
WM	0.38		0.18	0.33	1.32	0.58	0.31	1.31	0.47
E	0.22	0.18		0.18	0.84	0.59	0.22	1.15	0.41
M	0.14	0.33	0.18		1.00	0.49	0.39	1.09	0.59
F	0.98	1.32	0.84	1.00		1.63	0.72	2.17	1.09
C	0.33	0.58	0.59	0.49	1.63		0.62	0.29	0.27
S	0.40	0.31	0.22	0.39	0.72	0.62		1.17	0.45
N	0.76	1.31	1.15	1.09	2.17	0.29	1.17		0.45
V	0.35	0.47	0.41	0.59	1.09	0.27	0.45	0.45	

pairs of two parts are each roughly proportional. This can easily be explained by the fact that countries where veal (red meat) is available are also rich in milk, and the same happens with chicken (white meat) and eggs. On the other side of the spectrum, the largest contributions to variability are given by the logratios involving fish or nuts (the variance of the logratio of fish to nuts is 2.17, but all logratios involving fish are larger than 0.7, and the same can be said of almost all logratios of nuts, i.e., of all but 2).

These results are verified by a biplot (Figure 5.13), which represents 77% of the total variability. At first sight, one can define roughly three groups of parts: cereals–nuts–vegetables, whose relative consumption increases toward the left and slightly to the top of the diagram, meats–milk–eggs–starch increasing toward the lower-right corner of the diagram, and fish alone increasing toward the upper-right corner of the diagram. Note that these statements relate to the rays, thus to the clr-transformed variables: such directions of *increase* are actually directions of higher percentage of consumption relative to all parts considered here. The largest contributions are given by the rays of fish, nuts, and white meat, as already indicated by the variation matrix. Note that long rays pointing in different directions will necessarily give quite long links, thus large contributions to the total variance. Aligned vertices suggest one-dimensional patterns of subcompositions, to be further explored. For each such subcomposition, one should proceed to a reclosure, a clr transformation of the subcomposition, an SVD of the scores obtained, and the calculation of the proportion of variance represented by the first component with regard to the total variance of the subcomposition. In case of three-part subcompositions, a

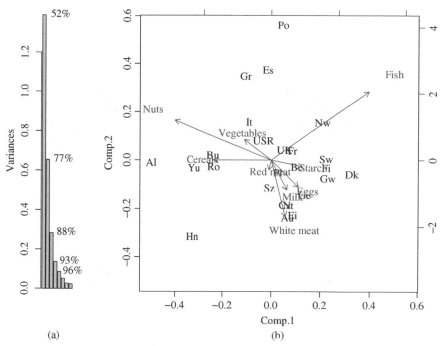

(a)   (b)

Figure 5.13   (a) Scree plot and (b) biplot of the nutrition data set. The scree plot reports the variance of each PC (bars) as well as their explained cumulative variance (in %). Note that the first two PCs explain less than 80% of the variance.

ternary diagram can also be used to display a simplicial line passing through the center of the subcomposition, and along the direction of its first PC. In this case, candidate alignments (and proportions of variance explained by the first PC) are

- white meat, eggs, and starch (67%);
- cereals, red meat, and starch (69%);
- eggs, red meat, and cereals (79%);
- red meat, milk, and cereals (81%);
- white meat, starch, and fish (87%);
- nuts, red meat, and eggs (88%);
- cereals, vegetables, and fish (90%);
- nuts, cereals, and white meat (92%).

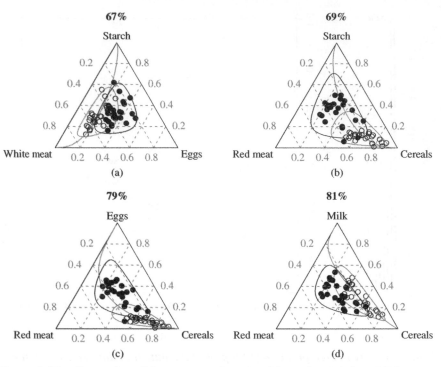

Figure 5.14   Examples of three-part subcompositions that might exhibit a one-dimensional pattern according to the biplot of Figure 5.13. Each diagram shows the original (circles) and centered (filled circles) data, the first PC for the original data, and elliptic 95%-probability regions (both ellipses have the same compositional shape, but different center). Each geometric element is in black for the centered data or gray for the original data. Diagram captions report the percentage of variance explained by this first component. Note that the larger this percentage, the more elongated the ellipses should be.

Figures 5.14 and 5.15 show all three-part subcompositions mentioned, with indication of the first PC and the percentage of variance within that subcomposition explained by it. In this way, the reader can get an idea of what the several percentages mean in terms of variability around the main trend given by the PC.

Table 5.8 reports the two PCs of one of these three-part subcompositions, formed by white meat, cereals, and nuts. The reader can check that the two vectors reported in the table are actually orthogonal, that is, their inner product

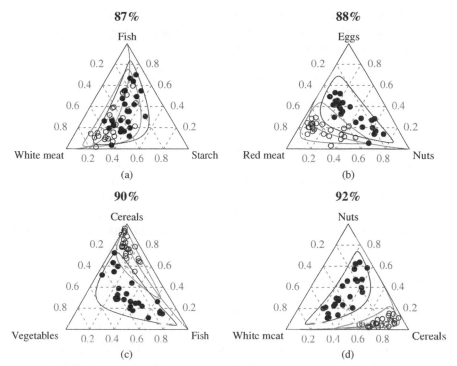

Figure 5.15   Examples of three-part subcompositions, continued from Figure 5.14.

Table 5.8   Principal components of the subcomposition white meat–cereals–nuts, expressed as clr components (note the zero sum of each row).

	White meat	Cereals	Nuts	λ_i
PC.1	0.758	−0.116	−0.642	0.816
PC.2	0.304	−0.808	0.504	0.244

is zero. From the definition of PCs, these two sets of numbers can be rewritten in the form of two equations

$$PC_1 = 0.758 \ln(\text{WM}) - 0.116 \ln(\text{C}) - 0.642 \ln(\text{N}), \tag{5.7}$$

$$PC_2 = 0.304 \ln(\text{WM}) - 0.808 \ln(\text{C}) + 0.504 \ln(\text{N}). \tag{5.8}$$

If the first one captures 90% of the variance, the second one must represent 10% of the variance and can be taken as almost constant. Taking exponentials and rescaling the coefficients by a factor of 10,

$$\exp(PC_2) \approx \exp(-10)\frac{(WM)^3(N)^5}{(C)^8} = k,$$

suggesting a sort of *standard recipe* of cereals combined with varying proportions of nuts to white meat: some countries use more white meat, others more nuts, to complement the widespread use of cereals. Note, however, that even the countries with a *higher* nut consumption eat more white meat than nuts: in the end, *higher consumption of A* means in a compositional context *higher average consumption of A relative to the other food sources*. Similar *recipes* can be extracted from the other three-part subcompositions. For instance, vegetables are complemented with varying proportions of fish to cereals; starch is mixed with fish or white meat; and red meat consumption gets balanced with varying mixtures of eggs and nuts.

With regard to the distribution of markers, it is possible to define three to four groups of countries. Western-block Mediterranean countries (Italy, Greece Spain, and also Portugal) are characterized by high relative consumptions of nuts, fish, vegetables, and cereals, but specially fish. North and Central countries (Scandinavian countries, both German states, Ireland, Belgium, Poland, Netherlands) are grouped by their high consumption of starch, eggs, and milk. Within this group, we can order the countries by their fish-to-white meat ratio from Norway (and Scandinavia as a whole) to Ireland–Switzerland–Austria (strangely, Ireland had a fish profile consumption similar to that of the two alpine countries). Eastern-block south countries (Yugoslavia, Hungary, Albania, Bulgaria, and Romania) have also a high relative consumption in nuts and vegetables, and very low relative consumption of protein sources (specially fish in Hungary). Finally, Russia, France, and United Kingdom (perhaps including Italy, Belgium, and Poland) are placed quite in the middle, showing a nutrition profile close to the average of European population, although for each country, this might be for different reasons (geographic centrality for France or United Kingdom, sheer size for USSR).

5.9 A political example

The Spanish territorial structure recognizes to each region an important level of autonomy, comparable to that of many federal states – such as Germany, Canada, or Australia – although its 1979 Constitution does not explicitly state that the country is a federal state. This apparent contradiction is related to the existence of different and sometimes contradictory feelings of *nation* and *nationality* within the country. That adds a bit of extra complexity to the typical labor-conservative distribution of political parties. In particular, in the autonomous region of Catalonia, parties must also define a profile in the matters of nationalism versus nonnationalism, and in Spanish versus Catalan nationalism. How do the citizens perceive the several parties? Is the nationalist matter more or less important than the social classical left–right division? To answer these questions and, in general, to get an election profile of Catalan society, a data set of votes cast to each party in the elections between 1980 and 2006 (general or regional) by electoral districts is analyzed. This might serve as an example for other regions in the world with a strong independence or reconfederation movement (Québec, Flanders, Scotland). Six parties are considered (by decreasing total number of votes):

1. Catalan Socialist Party – Spanish Labor-Socialist Party (PSC-PSOE), the historical social-democratic party in Spain; member of the European Socialist Party and the International Socialist organization;

2. Catalan Democratic Convergence (CDC) – Catalan Democratic Union (UDC), a long-standing coalition (CiU) of two parties; CDC is a member of the European Liberal Party, while UDC is a member of the European Popular Party;

3. Spanish People's Party (PP), the classical conservative Spanish party, also member of the European Popular Party; PP existed until 1989 with another name; also, a liberal party (CDS) participated in several elections, but it was finally absorbed by PP; therefore, in all elections, their votes have been amalgamated;

4. Catalonia's Republican Left party (ERC), a Catalan party proposing the independence of Catalonia from the rest of Spain; it claims to have a labor component, being different from PSC for its pro-independence profile;

5. Catalonia Initiative-Greens–United Alternative Left (IC), a varied coalition of several communist, socialist, and green parties, whose members belong to the European Communist Party or the European Green party; they took part in some elections as separate parties, although we have amalgamated their votes in all cases;

6. Catalonia's Citizens (C's or CC), a new party born after 2005, the first time in more than 20 years that CiU did not govern Catalonia (actually, PSC-PSOE, IC, and ERC formed a three-party government, which triggered the formation of this new group); given its short age, the social perception of this party is unclear, but they define themselves as social-democratic and nonnationalist. From a purely statistical point of view, the data set will have missing values for this formation before 2006, adding an extra level of complexity to the analysis.

Before going on, one should take into account that in the regional elections of 1999, IC and PSC-PSOE formed a coalition, while C's did not yet exist. This is seen in the data set as 0 votes for IC at this year, although the data are actually amalgamated into the PSC column. To solve this issue and undo the amalgamation, the replacement strategy of Section 2.3.2 is followed, considering that the logratio of votes of these two parties should have been consistent with the observed ratios in 1995 and 2003. For each shire, the following factor is calculated:

$$\kappa = \frac{1}{2} \left[\ln \frac{IC_{1995}}{PSC_{1995}} + \ln \frac{IC_{2003}}{PSC_{2003}} \right],$$

and then the reported votes PSC_{1999} for each shire are split among IC and PSC as follows:

$$IC^*_{1999} = \frac{e^\kappa}{1 + e^\kappa} \cdots PSC_{1999}, \quad PSC^*_{1999} = \frac{1}{1 + e^\kappa} \cdots PSC_{1999}.$$

These data are finally used for a standard biplot analysis. Four biplots are reported in Figure 5.16: a biplot of all the elections before 2006, and three biplots relating to the election of 2006, when the party C's took part. Before 2006, parties were distributed in a triangular shape with three poles: ERC, PP, and IC, the minor parties of the Catalan political system. This is not a surprise, as their smaller size implies that their relative changes are going to be larger, thus having a larger relative variance than the central parties (CiU and PSC). In this shape, PSC is placed between PP and IC, while CiU has an intermediate position between PP and ERC. Ternary diagrams of these two 3-party subcompositions (with percentage of variance explained by the first component) are reported in Figure 5.17. It can be seen that the splitting among IC, PSC, and PP can be mostly explained by a single effect, possibly the classical left–right dialectics. The alignment PP–CiU–ERC captures a lower proportion of variance but can be interpreted as a Spanish–Catalan nationalism axis. These two axes do not appear orthogonal before 2006, suggesting that these two components of Catalan politics were not independent from the voters' point of view. Moreover, the left–right dynamics (roughly aligned with the first PC) explained 45% of the variance, while the nationalistic component (associated to the second PC) accounted for an additional nonredundant 30% of the variance.

After the appearance of C's, the picture shows both important similarities and striking differences (comparing both biplots without C's). The general structure of parties remains, as well as their relative positions (except for the change of position of CiU), although their structure has rotated. In the elections of 2006, the first axis has become the Spanish–Catalan nationalistic confrontation (claiming as much as two-thirds of the variance), while the left–right classical distribution represented just one-sixth of the variability in the subcomposition without C's.

However, if we take into account the presence of C's, we see that the first axis is absolutely dominated by the variability of this party, which is responsible for up to 84% of the variability in the data. Clearly, C's is opposed to CiU and ERC in this first axis, suggesting that voters have seen C's as an

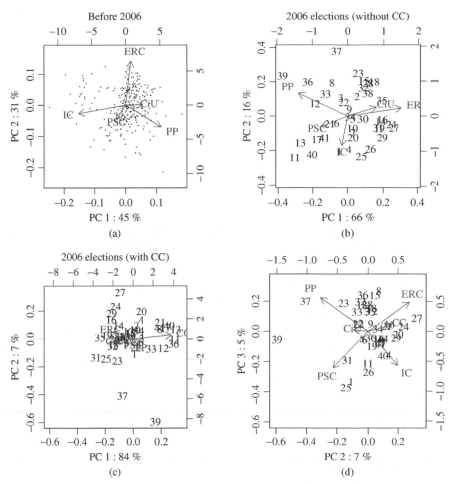

Figure 5.16 Biplots of the election data set, with indication of percentage of variance explained by each principal component (see axes). Panel (a) gives the biplot of all data before 2006, without the party CC. The remaining biplots relate to the data of the election of 2006: the panel (b) gives the biplot of the subcomposition without CC, while panels (c) and (d) are biplots of the whole composition, first versus second PCs (a,c) and second versus third PCs (b,d).

anti-Catalan-nationalistic movement (not necessarily pro-Spanish-nationalistic, as C's is not linked to PP). In contrast, PP–PSC–IC defines the second axis (the classical right–left alignment, representing 7% of the variability). Of a similar importance, the third PC (5% of the variance) shows a striking and yet unexplained structure, with ERC and PP on one side and IC and PSC on the other side: a look at the shires does not allow to set a relation of this third axis with

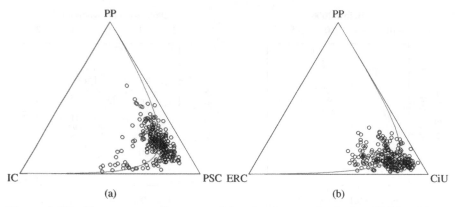

Figure 5.17   Two ternary diagrams of the election results before 2003, with indication of percentage of explained variance by the first principal component in each set. First PC explained variance: (a) 82.91 and (b) 65.09.

any of the classical sociological explanations (urban–rural, mountain–coast, populated–depopulated, younger–older).

As a summary, we may conclude that Catalan voters before 2006 identified two main axes of politics at the time of voting: first they decided in terms of labor-conservative ideas (axis IC–PSC–PP), then in terms of whether they felt more Spanish or more Catalan. In 2006, the situation had changed dramatically, and the first decision was in terms of pro-Catalan-nationalism or nonnationalism (CiU& ERC vs. C's), followed by far by the questions of left/independentist (ERC–IC)–right/unionist(PSC–PP) policies (second and third axes), which were totally blurred by the appearance of C's and its confrontation with CiU.

5.10   Exercises

Exercise 28. This exercise aims at illustrating the problems of classical statistics applied to compositional data. Using the data given in Table 5.1, compute the classical correlation coefficients between the following pairs of parts: (MnO vs. CaO), (FeO vs. Na_2O), (MgO vs. FeO), and (MgO vs. Fe_2O_3). Now ignore the structural oxides Al_2O_3 and SiO_2 from the data set, close the remaining components to percentages within the subcomposition, and compute the same correlation coefficients as above. Compare the results. Compare the correlation matrix between the feldspar-constituent parts (CaO,Na_2O,K_2O), as obtained from the original data set, and after closing this three-part subcomposition.

Exercise 29. For the data given in Table 2.1, compute and plot the center with the samples in a ternary diagram. Compute the total variance and the variation matrix.

Exercise 30. Perturb the data given in Table 2.1 with the inverse of the center. Compute the center of the perturbed data set and plot it with the samples in a ternary diagram. Compute the total variance and the variation matrix. Compare your results numerically and graphically with those obtained in Exercise 29.

Exercise 31. Make a biplot of the data given in Table 2.1 and give an interpretation.

Exercise 32. Figure 5.5 shows the biplot of the data given in Table 5.1. How would you interpret the different patterns that can be observed?

Exercise 33. Select three-part subcompositions that behave in a particular way in Figure 5.5 and plot them in a ternary diagram. Do they reproduce properties mentioned in the previous description?

Exercise 34. Do a scatter plot of the logratios

$$\frac{1}{\sqrt{2}} \ln \frac{K_2O}{MgO} \quad \text{against} \quad \frac{1}{\sqrt{2}} \ln \frac{Fe_2O_3}{FeO},$$

identifying each point. Compare with the biplot. Compute the total variance of the subcomposition $(K_2O, MgO, Fe_2O_3, FeO)$ and compare it with the total variance of the full data set.

Exercise 35. How would you recast the data in Table 5.1 from mass proportion of oxides (as they are) to molar proportions? You may need the following molar weights. Any idea of how to do that with a perturbation?

SiO_2	TiO_2	Al_2O_3	Fe_2O_3	FeO	MnO
60.085	79.899	101.961	159.692	71.846	70.937

MgO	CaO	Na_2O	K_2O	P_2O_5	
40.304	56.079	61.979	94.195	141.945	

Exercise 36. Redo all the descriptive analysis (and the related exercises) with the Kilauea data set (Table 5.1) expressed in molar proportions. Compare the results.

Exercise 37. Compute the vector of arithmetic means of the ilr-transformed data from Table 2.1. Apply the ilr^{-1} backtransformation and compare it with the center.

Exercise 38. Take the parts of the compositions in Table 2.1 in a different order. Compute the vector of arithmetic means of the ilr-transformed sample. Apply the ilr^{-1} backtransformation. Compare the result with the previous one.

Exercise 39. Center the data set of Table 2.1. Compute the vector of arithmetic means of the ilr-transformed data. What do you obtain?

Exercise 40. Compute the covariance matrix of the ilr-transformed data set of Table 2.1 before and after perturbation with the inverse of the center. Compare both matrices.

Exercise 41. Take the Catalan elections data set, and select the data for 2006 (e.g., filtering it in a spreadsheet software). Compute center, total variance, and variation matrix of the subcomposition (PSC, CiU, PP, ICV, ERC). Compute these statistics for the whole composition of parties (including C's with all the preceding ones). Compare the two variation matrices. Take the mean of the whole composition, remove the value for C's, reclose the remaining five means to 100% and compare with the mean originally computed for the subcomposition.

Exercise 42. Take the Catalan elections dataset, and select the data prior to 2006 (e.g., filtering it in a spreadsheet software). Compute center and total variance of the subcomposition (PSC, CiU, PP, ICV, ERC). Compare with the equivalent statistics obtained in the preceding exercise. Which was the perturbation occurring in an average within this subcomposition as a consequence of the appearance of C's? Think that *no change* would mean a perturbation of $[1, 1, 1, 1, 1]/5$. What happened with the variability? Compare by quotient.

Exercise 43. As in the preceding exercise, compute the two matrices of variation of the subcomposition (PSC, CiU, PP, ICV, ERC) before and after 2006. Compare each entry of the variation matrices by quotient. Which pairs notably increased or decreased their contribution to the total variability? Which seem to have an equivalent contribution, that is, which have a similar ratio $t_{ij}^{after}/t_{ij}^{before}$ as the total variance ratio calculated before?

Exercise 44. Draw a ternary diagram of the subcomposition (PP, PSC, ICV) using only data before 2006. Include the first PC. How much variability is captured by this first PC? Compare with the proportion of variance explained by the first component in the biplot of Figure 5.16a. Repeat the exercise with the subcomposition (PP, CiU, ERC), and compare with the second PC of the same biplot.

6

Random compositions

6.1 Sample space

Random variables, random functions, stochastic processes, and other more complex random objects, are mathematical models of experiments or observational procedures which appear in practice. For simplicity, all these cases are here called *random variables*. Mathematical models for random variables are made of, at least, three elements: a probability space, a sample space, and a function from the probability space to the sample space that is identified with the random variable. The *probability space* is an abstract representation of the possible conditions in which the experiment is carried out. The *sample space* contains the possible outputs of the experiment. The experiment itself is a function that assigns the output of the experiment to each experimental condition in the probability space. The mathematical details are summarized in Appendix B. In practice, attention is centered in the sample space and its structure. Some examples follow.

Example 6.1 (Wind speed measurement).
Consider an experiment that consists in measuring the horizontal wind speed at a given point. Assuming that the experiment is absolutely accurate and that the wind speed is unbounded, all positive real numbers are possible outputs of the experiment. The sample space of this experiment must contain the positive real

Modeling and Analysis of Compositional Data, First Edition.
Vera Pawlowsky-Glahn, Juan José Egozcue and Raimon Tolosana-Delgado.
© 2015 John Wiley & Sons, Ltd. Published 2015 by John Wiley & Sons, Ltd.

numbers, $\mathbb{R}_+$. However, other impossible outputs can be added to the sample space. For instance, the 0 wind speed is a physically impossible result of an absolutely accurate measurement; it can appear only as an *under detection limit* measurement. Negative wind speed is impossible by definition, but negative real numbers can be added to the sample space. This practice of adding impossible outputs of the experiment to the sample space is a common practice in probabilistic modeling, but it can lead to paradoxical situations, as commented in the next example. ◇

Example 6.2 (Individual preference).
In a survey, an individual is asked to select, among three options, a single preferred category; the individual may prefer not to give a response. The result of this kind of experiment is one out of four categories including nonresponse. If categories are labeled A_i, $i = 1, 2, 3, 4$, the output of the experiment is a pointer indicating which was the response of the individual. For instance, the four-position array $(0, 0, 1, 0)$ indicates that the individual response was A_3. Then, the sample space can be the set of four-position arrays, containing 0 in three positions and a 1 in the remaining position. Again this sample space can be augmented with nonpossible four-position arrays containing more than a single 1 or even other symbols different from 0 and 1. As in the previous example, this unnecessary practice may lead to some paradoxical situations. ◇

Example 6.3 (Preferences of n individuals).
Assume the previous experiment on individual preference is repeated on n different individuals. The results of interest are the number of individuals giving the same response. Now the experiment consists of recording the four numbers of responses in each category. Let n_i be the number of responses to A_i. The result can be coded as an array (n_1, n_2, n_3, n_4), where $n = n_1 + n_2 + n_3 + n_4$, $n_1 = 0, 1, 2, \ldots, n - n_2 - n_3 - n_4$, and analogously for n_i, $i = 2, 3, 4$. Therefore, the sample space can be the set of four-position arrays containing nonnegative integers adding up to n. Although this sample space recalls the four-part simplex normalized to $\kappa = n$, it is necessary to insist on the fact that zero counts, that is, $n_i = 0$, cannot be included in the simplex (Definition 2.4). ◇

Example 6.4 (Proportions of preferences).
A step further with the survey example: after surveying a large population of individuals, the experiment is now redefined as recording the proportions of individuals selecting each category. The output of the experiment is a composition of four parts, $[x_1, x_2, x_3, x_4]$, in S^4, where x_i is the proportion of individuals selecting A_i and, consequently, $x_1 + x_2 + x_3 + x_4 = 1$. In this case, the sample space must contain S^4 and, particularly, can be identified with S^4. The assumption that a

large population is surveyed for preferences has a hidden consequence: if any of the proportions is null, the corresponding category should be removed from the survey, as it is an impossible preference. Again, null or negative proportions can be added to the sample space at the price of subsequent incoherences. Also, note that observed frequencies are always rational numbers, which form a composition. Compositions with irrational parts are added to complete S^4.   ◇

When the result or output of an experiment is uncertain, the description of this uncertainty is a primary goal. This is done assigning probabilities to sets of possible experimental results called *events*. Thus, events are subsets of the sample space to which a probability is assigned. The set of all events is called σ-field of the sample space (see Appendix B for details). The σ-field is the elementary structure of the sample space that allows the development of a theory of random variables. When probabilities of all events in the σ-field are explicitly or implicitly specified, the random variable is completely described.

The probability of events of random variables with a sample space included in $\mathbb{R}^n$ is normally described using cumulative distribution functions (cdf) and/or probability density functions (pdf). There are different types of random variables: discrete, continuous, and mixed, depending on the characteristics of their pdf. The following discussion assumes continuous random variables. The main characteristic is that the pdf is a nonnegative integrable function defined on the sample space. When dealing with a random variable, $\mathbf{X} = [X_1, X_2, \ldots, X_D]$, taking values in the sample space S^D, some special features should be taken into account. At least, two different strategies are possible. The first one, here called *conventional approach*, considers S^D as a subset of $\mathbb{R}^{D-1}$ (or even a subset of $\mathbb{R}^D$). The second one, or *compositional approach*, considers S^D as a Euclidean space, which can be represented in coordinates that span the whole $\mathbb{R}^{D-1}$.

6.1.1   Conventional approach to the sample space of compositions

In the conventional approach, one part of the random composition $\mathbf{X} = [X_1, X_2, \ldots, X_D] \in S^D$, say $X_D = 1 - \sum_{i=1}^{D-1} X_i$, is removed from $\mathbf{X}$ to get the random vector

$$\mathbf{Y} = [X_1, X_2, \ldots, X_{D-1}] \in \mathbb{R}^{D-1}. \qquad (6.1)$$

Then, a pdf $f_{\mathbf{Y}}(\mathbf{y})$, with $\mathbf{y} \in \mathbb{R}^{D-1}$, is defined as follows:

$$P[\mathbf{Y} \in A] = \int_A f_{\mathbf{Y}}(\mathbf{y}) \; d\mathbf{y}, \qquad (6.2)$$

for any event A in the σ-field of $\mathbb{R}^{D-1}$, called Borelian sets of $\mathbb{R}^{D-1}$. However, as possible outputs of $\mathbf{X}$ are within S^D, the feasible values of $\mathbf{Y}$ are also restricted to satisfy the condition $\sum_{i=1}^{D-1} Y_i < 1$. Then, if A does not contain points satisfying the condition, the assumption that $P[\mathbf{Y} \in A] = 0$ is mandatory.

Some remarks are necessary to understand this strategy. To begin with, $d\mathbf{y}$ in Equation (6.2) indicates that the integral is carried out with respect to the Lebesgue measure in $\mathbb{R}^{D-1}$. This can be visualized approximating the integral in Equation (6.2) by a sum. Assume A to be a $(D-1)$-dimensional interval in $\mathbb{R}^{D-1}$ given by

$$A = \{[y_1, y_2, \dots, y_{D-1}] \mid \eta_{i0} \leq y_i < \eta_{ik}, \; i = 1, 2, \dots, D - 1\}.$$

The interval of each coordinate y_i is partitioned by the points $\eta_{i0} < \eta_{i1} < \dots < \eta_{ik}$. Then, under mild assumptions on the pdf $f_{\mathbf{Y}}$,

$$P[\mathbf{Y} \in A] \approx \sum_{j=0}^{k-1} f_{\mathbf{Y}}(\eta_{1j}, \eta_{2j}, \dots, \eta_{D-1,j}) \cdot \prod_{i=1}^{D-1} |\eta_{i,j+1} - \eta_{ij}|, \qquad (6.3)$$

where the product $\prod_{i=1}^{D-1} |\eta_{i,j+1} - \eta_{ij}|$ is the Lebesgue measure of the ith subinterval of the partition, that is, the conventional volume of a hyper-parallelepiped when the coordinates y_i are orthogonal.

A second remark concerns the cases in which A contains a single point of the sample space, known as *elementary event*. Consider $\mathbf{x}_0 \in S^D$ and $\mathbf{y}_0$, its reduced version in $\mathbb{R}^{D-1}$ (Equation 6.1); if A contains the single point $\mathbf{y}_0$, then, as implied by Equation (6.2), $P[\mathbf{X} = \mathbf{x}_0] = P[\mathbf{Y} = \mathbf{y}_0] = 0$, independently of the value of $f_{\mathbf{Y}}(\mathbf{y}_0)$. However, $\mathbf{x}_0$ is a possible output of the random composition $\mathbf{X}$. For comparison, consider $\mathbf{x}_1$ with the first component being negative. The reduced version $\mathbf{y}_1 \in \mathbb{R}^{D-1}$ still has the first component negative. It is clear that $P[\mathbf{X} = \mathbf{x}_1] = P[\mathbf{Y} = \mathbf{y}_1] = 0$. The paradoxical fact is that $P[\mathbf{X} = \mathbf{x}_0] = P[\mathbf{X} = \mathbf{x}_1] = 0$, but $\{\mathbf{x}_0\}$ is a possible event, while $\{\mathbf{x}_1\}$ is an impossible event, that is, probability is not able to distinguish between a possible elementary event and an impossible elementary event. The impossible events were artificially added to the sample space, and this fact causes the paradoxical result.

6.1.2 A compositional approach to the sample space of compositions

The representation of compositions in coordinates provides a straightforward way of defining the sample space of a random composition. Let $\mathbf{X} = [X_1, X_2, \dots, X_D]$ be a random composition taking values in S^D, that is, S^D is included in the sample space. Select some orthonormal basis of the simplex with its corresponding contrast matrix $\boldsymbol{\Psi}$ (see Section 4.4, Definition 4.3). The ilr coordinates associated with this basis are (Equation 4.7)

$$\mathbf{X}^* = [X_1^*, X_2^*, \ldots, X_{D-1}^*] = \mathrm{ilr}(\mathbf{X}) = \ln(\mathbf{X})\ \boldsymbol{\Psi}^\top.$$

The vector of coordinates $\mathbf{X}^*$ is a random variable, in fact a random vector, with sample space $\mathbb{R}^{D-1}$. The standard σ-field in $\mathbb{R}^{D-1}$ can be identified with the Borelian sets (Appendix B). A subset in the σ-field of S^D is the ilr^{-1} image of a Borelian set of $\mathbb{R}^{D-1}$. For instance, if A^* is an interval in $\mathbb{R}^{D-1}$, $A = \mathrm{ilr}^{-1}(A^*)$ is an interval in S^D and an element of the σ-field. In order to specify the probabilities of each set in the σ-field of S^D, one can proceed in the following way: define a pdf for the random coordinates $\mathbf{X}^*$, denoted as $f^*(\mathbf{x}^*)$, such that, for any A^* in the σ-field of $\mathbb{R}^{D-1}$,

$$P[\mathbf{X}^* \in A^*] = \int_{A^*} f^*(\mathbf{x}^*)\ d\mathbf{x}^*.$$

The probability of $A = \mathrm{ilr}^{-1}(A^*)$ in the σ-field of S^D is simply assigned to be $P[\mathbf{X} \in A] = P[\mathbf{X}^* \in \mathbf{A}^*]$. Summarizing, the sample space of a random composition is the simplex S^D; the σ-field is defined as the inverse ilr of the Borelians in $\mathbb{R}^{D-1}$; and the probabilities are transferred from the Borelians in $\mathbb{R}^{D-1}$ to their ilr^{-1} images in S^D. It can be proven that the probability assigned to any A in the σ-field of S^D does not depend on the selected basis and its associated ilr transformation.

6.1.3   Definitions related to random compositions

The following definitions summarize the previous content of this section.

Definition 6.5 (Sample space of a random variable).
The sample space of a random variable X is a set containing all possible values of X. It is necessary to specify a σ-field of subsets of the sample space.

The previous definition matches the standard probability theory summarized in Appendix B. In practice, sample spaces have a richer structure than that implied in the σ-field, although, many times, it is only implicitly assumed. For instance, the study of real random variables, with sample space $\mathbb{R}^n$, implicitly assumes the Euclidean structure of $\mathbb{R}^n$, with the usual operations (sum and product), and metrics (distance, inner product, Lebesgue measure). When dealing with random compositions, it is relevant to specify which are the operations and metric assumed in the simplex as sample space. As presented in Chapter 3, S^D has a Euclidean structure. The following definition assumes that the Aitchison geometry of the simplex is a proper structure of S^D.

Definition 6.6 (Random composition).
The random variable **X** *is a D-part random composition if all its possible values are in* S^D, *which is taken as the sample space. Moreover, simplicial operations (perturbation and powering) and metrics (Aitchison inner product, norm, and distance) are assumed valid in the sample space* S^D.

The assumption that the Aitchison geometry is the proper structure of the sample space allows to think of perturbation of two random compositions. If **X** and **Y** are random compositions, $\mathbf{Z} = \mathbf{X} \oplus \mathbf{Y}$ is also a random composition. Each output of **Z** is the perturbation of the respective outputs of **X** and **Y**. As the outputs of **Z** are in S^D, it is the sample space of **Z**. Similarly, for a real constant α, $\mathbf{Z} = \alpha \odot \mathbf{X}$ is a random composition. The Aitchison norm of **X**, $\|\mathbf{X}\|_a$ is a random variable the nonnegative real numbers as sample space. Similar statements are valid for the Aitchison inner product of random compositions or for their distance.

In practice, most random compositions are assumed to be continuous, that is, their probability distribution is absolutely continuous (see Appendix B). Continuous random compositions are defined as follows.

Definition 6.7 (Continuous random composition. Probability density function).
A random composition, **X**, *in* S^D, *is continuous if there exists a nonnegative real function* $f^* : \mathbb{R}^{D-1} \to \mathbb{R}$, *defined almost everywhere in* $\mathbb{R}^{D-1}$, *such that for any Borelian set B in* $\mathbb{R}^{D-1}$,

$$P[\mathrm{ilr}(\mathbf{X}) \in B] = \int_B f^*(\mathbf{x}^*) \, d\mathbf{x}^*.$$

If f^* *exists, it is called pdf of the* ilr-*coordinates. The compound function* $f :$ $S^D \to \mathbb{R}$, *defined as* $f(\mathbf{x}) = f^*(\mathrm{ilr}(\mathbf{x}))$ *for any* $\mathbf{x} \in S^D$, *is called pdf of* **X** *on the simplex.*

The expression of the pdf of the ilr-coordinates depends on the particular orthonormal basis selected in the simplex to compute them. However, the pdf on the simplex does not depend on the choice of the orthonormal basis.

6.2 Variability and center

A continuous random composition is fully described by its pdf on the simplex or the pdf of its coordinates in a given basis of the simplex. However, summaries of the probability distribution are useful to understand the main characteristics. First- and second-order moments are frequently used for describing real random variables, and these concepts need to be translated for random compositions. To proceed, the ideas introduced in Fréchet (1948) are applied to define variability and expectation in the simplex (Egozcue and Pawlowsky-Glahn, 2001, 2002; Pawlowsky-Glahn and Egozcue, 2011). The key point is that in the sample space, a distance is defined. For random compositions, the Aitchison distance is adopted.

Definition 6.8 (Variability of a random composition).
Let $\mathbf{X}$ *be a D-part random composition and* $\mathbf{z}$ *a composition in* S^D. *The variability of* $\mathbf{X}$ *with respect to* $\mathbf{z}$ *is*

$$\text{var}[\mathbf{X}; \mathbf{z}] = E[d_a^2(\mathbf{X}, \mathbf{z})],$$

where the expectation $E[\cdot]$, *if it exists, is the expectation of a real random variable.*

Definition 6.9 (Center and total variance).
Let $\mathbf{X}$ *be a D-part random composition in which variability exists. The expectation in* S^D *or center of* $\mathbf{X}$ *is*

$$\text{cen}[\mathbf{X}] = \underset{z \in S^D}{\text{argmin}} \left\{ \text{var}[\mathbf{X}; \mathbf{z}] \right\},$$

and the minimum value of variability attained is the total variance

$$\text{totvar}[\mathbf{X}] = \underset{z \in S^D}{\min} \left\{ \text{var}[\mathbf{X}; \mathbf{z}] \right\}.$$

The effective computation of the center and total variance is commonly carried out using alternative formulas similar to the mean and variance of real variables. Next theorem gives alternative expressions for the center of a random composition (Pawlowsky-Glahn and Egozcue, 2001; Egozcue and Pawlowsky-Glahn, 2011).

Theorem 6.10 *Let* $\mathbf{X}$ *be a D-part random composition in which variability exists. Then,*

$$\text{cen}[\mathbf{X}] = \text{ilr}^{-1}(E[\text{ilr}(\mathbf{X})]) = \text{clr}^{-1}(E[\text{clr}(\mathbf{X})]) = C \exp(E[\ln \mathbf{X}]),$$

where ilr *is taken with respect to an arbitrary orthonormal basis.*

The total variance of a random composition $\mathbf{X}$ can be decomposed into variances of ilr-coordinates, clr coefficients and simple logratios of parts. The next theorem accounts for these decompositions and gives useful expressions to compute the total variance (Egozcue and Pawlowsky-Glahn, 2011).

Theorem 6.11 *Let* **X** *be a D-part random composition in which variability exists. Consider an arbitrary orthonormal basis of* S^D *and the corresponding random ilr-coordinates of* **X**, *denoted by* X_i^*, $i = 1, 2, \ldots, D-1$. *Denote* clr(**X**) = [clr$_1$(**X**), clr$_2$(**X**), $\ldots$, clr$_D$(**X**)]. *Then,*

$$\text{totvar}[\mathbf{X}] = \sum_{i=1}^{D-1} \text{var}[X_i^*] = \sum_{j=1}^{D} \text{var}[\text{clr}_j(\mathbf{X})] = \frac{1}{2D} \sum_{i=1}^{D} \sum_{j=1}^{D} \text{var}\left[\ln \frac{X_i}{X_j}\right]. \quad (6.4)$$

The decomposition of the total variance (Equation 6.4) into variances of ilr-coordinates corresponds to the expression of **X** in ilr-coordinates (Equation 4.7),

$$\mathbf{X} = \bigoplus_{i=1}^{D-1} X_i^* \odot \mathbf{e}_i, \quad X_i^* = \langle \mathbf{X}, \mathbf{e}_i \rangle_a,$$

where the $\mathbf{e}_i$, $i = 1, 2, \ldots, D-1$, are the elements of the orthonormal basis of the simplex used for the ilr-coordinates. Moreover, the decomposition into variances of logratios and variances of clr coefficients recalls the definition of Aitchison distance between **X** and its center cen[**X**] (Definitions 3.7 and 6.8).

The total variance is a measure of dispersion of the random variable around the center. To support this assertion, an inequality of Chebyshev-type is useful and is accounted for in the next theorem.

Theorem 6.12   **(Chebyshev inequality).**
Let **X** *be a D-part random composition in which total variance,* totvar[**X**], *exists, and with center* cen[**X**]. *For any* $r > 0$,

$$P[d_a(\mathbf{X}, \text{cen}[\mathbf{X}]) \le r\sqrt{\text{totvar}[\mathbf{X}]} \ ] \ge 1 - \frac{1}{r^2}.$$

Proof. For any ilr-coordinates, consider the exterior of a circle of radius r around $\boldsymbol{\mu}^* = $ ilr(cen[**X**]),

$$A^* = \left\{ \mathbf{x}^* \in \mathbb{R}^{D-1} \,\middle|\, \sum_i (x_i^* - \mu_i^*)^2 \ge r^2 \right\}.$$

The total variance in Definition 6.9, expressed as an integral, is bounded, as

$$\text{totvar}[\mathbf{X}] = \int_{\mathbb{R}^{D-1}} d^2(\mathbf{x}^*, \boldsymbol{\mu}^*) f^*(\mathbf{x}^*) \, d\mathbf{x}^*$$

$$\ge \int_{A^*} d^2(\mathbf{x}^*, \boldsymbol{\mu}^*) f^*(\mathbf{x}^*) \, d\mathbf{x}^*$$

$$= \int_{A^*} \sum_{i=1}^{D-1} (x_i^* - \mu_i^*)^2 f^*(\mathbf{x}^*) \, d\mathbf{x}^*$$

$$\geq r^2 \int_{A^*} f^*(\mathbf{x}^*) \, d\mathbf{x}^* = r^2 \, P[\mathrm{ilr}(\mathbf{X}) \in A^*] = r^2 \, P[\mathbf{X} \in A],$$

where $A = \mathrm{ilr}^{-1}(A^*)$, which is the exterior of a circular set in S^D. Isolating $P[\mathbf{X} \in A]$ and setting $k = r^2$, the statement is proven.   ∎

The Chebyshev inequality in Theorem 6.12 confirms the total variance, or its square root, as a measure of dispersion. This measure represents a bound for the probability of events located far away from the center of the distribution.

In order to standardize the total variance when used for different dimensions, the following measure of dispersion is proposed.

Definition 6.13 (Simplicial standard deviation).
Let $\mathbf{X}$ *be a D-part random composition with defined total variance* totvar[$\mathbf{X}$] *and center* cen[$\mathbf{X}$]. *The parameter*

$$\mathrm{sstd}[\mathbf{X}] = \sqrt{\frac{\mathrm{totvar}[\mathbf{X}]}{D-1}}$$

is called simplicial standard deviation.

A similar normalized measure of dispersion was proposed in Graf (2005) and therein called variation coefficient. The sample version of the simplicial standard deviation, under the name metric standard deviation, was introduced in Boogaart and Tolosana-Delgado (2013). The Chebyshev inequality in Theorem 6.12 is rewritten in terms of the simplicial standard deviation as

$$P[d_a(\mathbf{X}, \mathrm{cen}[\mathbf{X}]) \geq k \cdot \mathrm{sstd}[\mathbf{X}]] \leq \frac{D-1}{k^2}, \quad k > 0,$$

where the dimension of the random composition appears explicitly.

Random compositions are frequently centered and/or standardized. These operations are specially important in exploratory analysis (Section 5.3). They are parallel to centering and standardization of real random variables, but some particularities appear in standardization.

Definition 6.14 (Centered random composition).
Let $\mathbf{X}$ *be a random composition with center* cen[$\mathbf{X}$]. *The operation*

$$\mathbf{Z} = \mathbf{X} \ominus \mathrm{cen}[\mathbf{X}]$$

is called the centering of $\mathbf{X}$.

Definition 6.15 (standardized random composition).
Let $\mathbf{X}$ *be a random composition with center* cen[$\mathbf{X}$] *and total variance* totvar[$\mathbf{X}$]. *The random composition*

$$\mathbf{Z} = \frac{1}{\sqrt{\text{totvar}[\mathbf{X}]}} \odot (\mathbf{X} \ominus \text{cen}[\mathbf{X}])$$

is called the standardization of $\mathbf{X}$.

Note that the total variance of a standardized random composition is 1 and the center is the neutral element of the simplex.

After the definition of standardization, one may wonder why standardization consists only of scaling by the constant $\sqrt{\text{totvar}[\mathbf{X}]}$, equal for all components, in contrast with common practice in real multivariate analysis, where each component is standardized separately or, alternatively, the real random vector is multiplied by the inverse of the covariance matrix, a practice also related to the Mahalanobis distance (Mahalanobis, 1936). The reason is that the scales of all parts of a random composition are linked together, and an unequal scaling of the parts destroys information inherent to the composition.

6.3   Probability distributions on the simplex

In most statistical analysis, probability models for random variables need to be specified. For statistical analysis of compositional data, some standard probability laws are available. The main distribution for compositional data is the normal on the simplex (Mateu-Figueras et al., 2013), also called additive logistic-normal distribution (Aitchison and Shen, 1980; Aitchison, 1982). A very popular, although rigid, distribution is the Dirichlet distribution (Aitchison, 1986; Narayanan, 1991) and its variants, such as the multivariate beta-distributions or compound distributions (Connor and Mosimann, 1969; Mosimann, 1962; Nadarajah and Kotz, 2007).

Probability distributions for random compositions are presented as pdf's. However, the expression of the pdf depends on the reference measure in the sample space, that is, how areas or volumes are measured in the sample space. When it is considered embedded in a real space, the Euclidean structure of the real space is implicitly inherited by the sample space, and the reference measure is the Lebesgue measure. Alternatively, when the sample space is the simplex with the Aitchison geometry, the reference measure should be the Aitchison measure. The adoption of these alternative measures produces different expressions of the corresponding pdf, although they represent the same probability distribution.

Before presenting the pdf expressions of the standard probability distributions, it is convenient to introduce some elements that characterize both Lebesgue and

Aitchison measures. The Lebesgue measure of an arbitrary interval in $\mathbb{R}^{D-1}$ characterizes the measure of all Borelian sets in $\mathbb{R}^{D-1}$ (Ash, 1972). This allows to restrict the definition to intervals as follows.

Definition 6.16 (Lebesgue measure of an interval).
Consider an arbitrary open interval in $\mathbb{R}^{D-1}$,

$$B = \{[y_1, y_2, \ldots, y_{D-1}] \in \mathbb{R}^{D-1} \mid a_i < y_i < b_i, \quad i = 1, 2, \ldots, D-1\}.$$

The Lebesgue measure of B is $\lambda(B) = \prod_{i=1}^{D-1} |b_i - a_i|$.

In Definition 6.16, the area of a rectangle for $D - 1 = 2$, or the volume of a parallelepiped for $D - 1 = 3$, are easily identified. They correspond to the ordinary concept of two- and three-dimensional real Euclidean spaces. The interval B is a Borelian set in $\mathbb{R}^{D-1}$, and the Lebesgue measure on open intervals is extended consistently to the whole σ-field of Borelian sets (see Appendix B).

The Aitchison measure in S^D is constructed from the Lebesgue measure on the space of ilr-coordinates.

Definition 6.17 (Aitchison measure of an interval).
Consider an orthonormal basis of S^D and its corresponding ilr *transformation into coordinates. Let B be the interval in $\mathbb{R}^{D-1}$ referenced in Definition 6.16 and the corresponding interval in S^D, $A = \text{ilr}^{-1}(B)$. The Aitchison measure of A is $\lambda_a(A) = \lambda(B)$, where $\lambda(B)$ is the Lebesgue measure of B.*

A pdf is a nonnegative function that, when integrated on an interval, results in the probability of that interval, as stated in Definition 6.7. Integrals are approximated by sums, as sketched in Equation (6.3) for an integral of a pdf with respect to the Lebesgue measure in $\mathbb{R}^{D-1}$. Consider now that $f^*(\mathbf{x}^*)$ is the pdf of the coordinates of a random composition $\mathbf{X}$ and the pdf $f(\mathbf{x}) = f^*(\text{ilr}(\mathbf{x}))$ as in Definition 6.7. The interpretation of f as a pdf depends on which type of integral of f produces the desired probabilities. These integrals, called Radon integrals with respect to a measure, can be approximated by sums such as

$$P[\mathbf{X} \in A] \approx \sum_j f(\mathbf{x}_j) \cdot \lambda_a(A_j),$$

where the intervals A_j constitute a partition of A and $\mathbf{x}_j$ is a point in A_j. This means that probabilities are approximately computed by summing products of values of the pdf f multiplied by the Aitchison measures of the corresponding subinterval. The density f is a pdf with respect to the Aitchison measure in S^D, while f^* is the pdf of the same probability distribution with respect to the Lebesgue measure of the ilr-coordinates.

Most concepts of random variables, as joint distribution or independence, apply to random coordinates of random compositions following the principle of working on coordinates (see Section 4.6 or Mateu-Figueras et al. (2011)). For instance, two random compositions are independent if their coordinates are independent real random vectors.

6.3.1   The normal distribution on the simplex

The *additive logistic-normal distribution* was introduced in Aitchison and Shen (1980) (see also Aitchison (1982, 1986)). The basic idea was to represent the random composition in additive logratio (alr) coordinates and then to assume that these coordinates follow a multivariate normal distribution. The corresponding density was expressed with respect to the Lebesgue measure on S^D. The same probability distribution is obtained if the ilr coordinates are assumed to follow a multivariate normal distribution, with the advantage that ilr-coordinates correspond to an orthonormal basis of the simplex. For this reason, the name *additive logistic-normal distribution* was simplified to *logistic-normal distribution*. Moreover, after Mateu-Figueras (2003), the pdf of the distribution was expressed with respect to the Aitchison measure in S^D (Mateu-Figueras et al., 2013), thus suggesting the name of *normal distribution on the simplex*. However, the three names correspond to the same probability law, although using a different parameterization and a different reference measure. In what follows, the ilr transformation is used to introduce the normal distribution on the simplex with the corresponding parameters.

Definition 6.18   (Normal distribution on the simplex).
Given a random composition $\mathbf{X}$, *with sample space* S^D, $\mathbf{X}$ *is said to follow a normal distribution on* S^D *if the vector of random orthonormal coordinates,* $\mathbf{X}^* = \mathrm{ilr}(\mathbf{X})$, *follows a multivariate normal distribution on* $\mathbb{R}^{D-1}$.

Therefore, if $\mathbf{X}$ follows a normal distribution on the simplex, $\mathbf{X}^*$ is multivariate normally distributed on $\mathbb{R}^{D-1}$, $\mathbf{X}^* \sim \mathcal{N}(\boldsymbol{\mu}, \boldsymbol{\Sigma})$, which pdf with respect to the Lebesgue measure in $\mathbb{R}^{D-1}$ is

$$f^*(\mathbf{x}^*) = \frac{1}{(2\pi)^{(D-1)/2} \, |\boldsymbol{\Sigma}|^{1/2}} \cdot \exp\left[-\frac{1}{2}(\mathbf{x}^* - \boldsymbol{\mu})\boldsymbol{\Sigma}^{-1}(\mathbf{x}^* - \boldsymbol{\mu})^\top\right], \qquad (6.5)$$

where the parameters $\boldsymbol{\mu}$ and $\boldsymbol{\Sigma}$ are the mean of the random ilr-coordinates and their covariance matrix, respectively. The covariance matrix $\boldsymbol{\Sigma}$ is frequently assumed nonsingular, that is, positive definite, so that $|\boldsymbol{\Sigma}| \neq 0$ and $\boldsymbol{\Sigma}^{-1}$ exists. The parameters of the normal distribution of the coordinates are taken as the parameters of the normal in the simplex for $\mathbf{X}$, and it is denoted by

$\mathbf{X} \sim \mathcal{N}_S(\boldsymbol{\mu}, \boldsymbol{\Sigma})$. According to Definition 6.7, the corresponding pdf with respect to the Aitchison measure in S^D is

$$f(\mathbf{x}) = \frac{1}{(2\pi)^{(D-1)/2} \, |\boldsymbol{\Sigma}|^{1/2}} \cdot \exp\left[-\frac{1}{2}(\mathrm{ilr}(\mathbf{x}) - \boldsymbol{\mu})\boldsymbol{\Sigma}^{-1}(\mathrm{ilr}(\mathbf{x}) - \boldsymbol{\mu})^\top\right], \quad (6.6)$$

where $\mathbf{x} \in S^D$ (Mateu-Figueras et al., 2013). It should be noted that the values of the parameters $\boldsymbol{\mu}$ and $\boldsymbol{\Sigma}$ depend on the particular selected ilr-transformation. However, the parameters change in a well-controlled way leaving the probability distribution invariant.

Theorem 6.19. *Consider two orthonormal basis in S^D with associated contrast matrices $\boldsymbol{\Psi}_0$ and $\boldsymbol{\Psi}_1$, such that the corresponding ilr-transformations are $\mathrm{ilr}_i(\mathbf{x}) = \ln \mathbf{x}\, \boldsymbol{\Psi}_i^\top$, $i = 0, 1$. Let $\mathbf{X}$ be a random composition normally distributed on the simplex, $\mathbf{X} \sim \mathcal{N}_S(\boldsymbol{\mu}_0, \boldsymbol{\Sigma}_0)$, where the parameters correspond to the pdfs in Equations (6.6) and (6.5) with the ilr-transformation ilr_0. Then, the coordinates of $\mathbf{X}$ obtained using ilr_1 follow a normal distribution on $\mathbb{R}^{D-1}$, $\mathcal{N}(\boldsymbol{\mu}_1, \boldsymbol{\Sigma}_1)$, where $\boldsymbol{\mu}_1 = \boldsymbol{\mu}_0\boldsymbol{\Psi}_0\boldsymbol{\Psi}_1^\top$ and $\boldsymbol{\Sigma}_1 = \boldsymbol{\Psi}_1\boldsymbol{\Psi}_0^\top\boldsymbol{\Sigma}_0\boldsymbol{\Psi}_0\boldsymbol{\Psi}_1^\top$ and, accordingly, $\mathbf{X} \sim \mathcal{N}_S(\boldsymbol{\mu}_1, \boldsymbol{\Sigma}_1)$ when the reference coordinates are obtained using ilr_1.*

Proof. Substitute in Equation (6.6), the argument $\mathrm{ilr}(\mathbf{x})$ by $\mathrm{ilr}_0(\mathbf{x}) = \mathrm{ilr}_1(\mathbf{x})\boldsymbol{\Psi}_1\boldsymbol{\Psi}_0^\top$. Group the $\boldsymbol{\Psi}$-matrices around $\boldsymbol{\Sigma}_0^{-1}$, and define

$$\boldsymbol{\mu}_1 = \boldsymbol{\mu}_0\boldsymbol{\Psi}_0\boldsymbol{\Psi}_1^\top, \quad \boldsymbol{\Sigma}_1^{-1} = \boldsymbol{\Psi}_1\boldsymbol{\Psi}_0^\top\boldsymbol{\Sigma}_0^{-1}\boldsymbol{\Psi}_0\boldsymbol{\Psi}_1^\top.$$

This expression matches that of a normal distribution on the simplex with mean $\boldsymbol{\mu}_1 = \boldsymbol{\mu}_0\boldsymbol{\Psi}_0\boldsymbol{\Psi}_1^\top$ and inverse covariance matrix $\boldsymbol{\Sigma}_1^{-1}$. To prove that $\boldsymbol{\Sigma}_1 = \boldsymbol{\Psi}_1\boldsymbol{\Psi}_0^\top \boldsymbol{\Sigma}_0\boldsymbol{\Psi}_0\boldsymbol{\Psi}_1^\top$ take into account the statement of Exercise 26. ∎

Two different pdfs for the normal on the simplex have been presented: the pdf, f^*, of coordinates in $\mathbb{R}^{D-1}$, with respect to the Lebesgue measure (Equation 6.5), and the pdf on S^D, f, with respect to the Aitchison measure (Equation 6.6). However, representing the pdf on the simplex with respect to the Lebesgue measure on S^D is common practice. This third pdf, now denoted by f_L for Lebesgue reference measure, is shown to be $f_L(\mathbf{x}) = f(\mathbf{x})/J(\mathbf{x})$, where $J(\mathbf{x}) = \sqrt{D} \prod_{i=1}^{D} x_i$ is the Jacobian of the transformation from $\mathbb{R}^{D-1}$ to S^D maintaining the Lebesgue reference measure (Mateu-Figueras et al., 2013). The difference between the two pdfs, f and f_L, of a normal on the simplex becomes apparent when the contours of the pdfs are compared. Figure 6.1 shows the contour plots of $f(\mathbf{x})$ (Figure 6.1a) and $f_L(\mathbf{x})$ (Figure 6.1b) in S^3 for parameters

$$\boldsymbol{\mu} = [0.3, 0.1], \quad \boldsymbol{\Sigma} = \begin{pmatrix} 1.2 & 0 \\ 0 & 1.2 \end{pmatrix}, \quad \mathrm{cen}[\mathbf{X}] = [0.255, 0.389, 0.356],$$

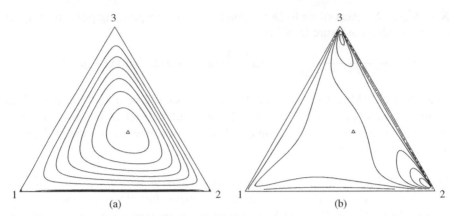

Figure 6.1   Contours of probability density functions of a normal random composition. The triangular symbol represents the center of the random composition. Parameters are specified in the text. (a) pdf with respect to the Aitchison measure. (b) pdf with respect to the Lebesgue measure.

where μ and Σ are referred to the ilr-coordinates

$$x_1^* = \sqrt{\frac{1}{2}}\ln\frac{x_2}{x_1}, \quad x_2^* = \sqrt{\frac{2}{3}}\ln\frac{x_3}{\sqrt{x_1 x_2}}.$$

The most striking difference in Figure 6.1 is that $f(\mathbf{x})$ (Figure 6.1a) shows a single mode, as expected for a normal distribution, but $f_L(\mathbf{x})$ appears with, at least, two modes, what is counterintuitive. In fact, a density $f_L(\mathbf{x})$ of a normal in the S^3 can show from 1 up to 3 modes (relative maxima) depending on both μ and Σ. The larger the variance of the coordinates is, the higher the chances of a multiple mode.

The parameter μ is the mean of the coordinates of a random composition. The center of a random composition $\mathbf{X} \sim \mathcal{N}_S(\mu, \Sigma)$ is computed as in Theorem 6.10,

$$\mathrm{cen}[\mathbf{X}] = \mathrm{ilr}^{-1}(\mathrm{E}[\mathrm{ilr}(\mathbf{X})]) = \mathrm{ilr}^{-1}(\mu),$$

that is, the center is the ilr^{-1}-transformed value of μ.

The importance of the normal distribution in the simplex is based on its properties of invariance and its asymptotic behavior. They are parallel to the properties of the real multivariate normal (see, e.g., Eaton, 1983; Shao, 1998), as they are obtained by an ilr^{-1}-transformation from the real space to the simplex. As stated in the next theorems, the normal distribution on the simplex is preserved under simplicial linear transformations of random compositions and under extraction of subcompositions.

Theorem 6.20. *Let* $\mathbf{X} \sim \mathcal{N}_S(\boldsymbol{\mu}, \boldsymbol{\Sigma})$ *be a random composition in* S^D. *If* β *is a real constant and* $\mathbf{a} \in S^D$, *then* $\mathbf{Y} = \mathbf{a} \oplus \beta \odot \mathbf{X}$ *is a random composition with normal distribution in* S^D, $\mathbf{Y} \sim \mathcal{N}_S(\boldsymbol{\mu}_Y, \boldsymbol{\Sigma}_Y)$, *with*

$$\boldsymbol{\mu}_Y = \text{ilr}(\mathbf{a}) + \beta \ \boldsymbol{\mu}, \quad \boldsymbol{\Sigma}_Y = \beta^2 \ \boldsymbol{\Sigma},$$

and $\text{cen}[\mathbf{Y}] = \mathbf{a} \oplus \beta \odot \text{cen}[\mathbf{X}]$.

Theorem 6.21. *Let* $\{\mathbf{X}_k \sim \mathcal{N}_S(\boldsymbol{\mu}_k, \boldsymbol{\Sigma}_k), k = 1, 2, \ldots, n\}$ *be a sequence of independent random compositions in* S^D. *Then, the perturbation* $\mathbf{Y} = \bigoplus_k \mathbf{X}_k$ *is a random composition with normal distribution in* S^D, $\mathbf{Y} \sim \mathcal{N}_S(\boldsymbol{\mu}_Y, \boldsymbol{\Sigma}_Y)$, *with*

$$\boldsymbol{\mu}_Y = \sum_{k=1}^n \boldsymbol{\mu}_k, \quad \boldsymbol{\Sigma}_Y = \sum_{k=1}^n \boldsymbol{\Sigma}_k,$$

and $\text{cen}[\mathbf{Y}] = \bigoplus_{k=1}^n \text{cen}[\mathbf{X}_k]$.

Theorem 6.22. *Let* $\mathbf{X} \sim \mathcal{N}_S(\boldsymbol{\mu}, \boldsymbol{\Sigma})$ *be a random composition in* S^D. *Let* $Sub(\cdot)$ *be an application that selects some indices of a composition. Thus, a random subcomposition* $\mathbf{X}_S$ *(after Definition 2.5) is built from a random composition* $\mathbf{X}$ *through* $\mathbf{X}_S = CSub(\mathbf{X})$. *The random subcomposition* $\mathbf{X}_S$ *follows then a normal distribution on the simplex with mean vector* $\boldsymbol{\mu}_S = CSub(\boldsymbol{\mu})$.

Averages of random compositions frequently appear as estimators of the center of a random composition, as well as in many applied situations. Consider a sequence of random compositions $\mathbf{X}_1, \mathbf{X}_2, \ldots, \mathbf{X}_n, \ldots$ The average of n of them in the simplex is

$$\overline{X}_n^* = \frac{1}{n} \odot \bigoplus_{i=1}^n \mathbf{X}_i.$$

The ilr-transformation into coordinates translates $\overline{\mathbf{X}}_n$ into the ordinary average of the coordinates

$$\overline{X}^* = \text{ilr}(\overline{X}_n) = \frac{1}{n} \sum_{i=1}^n \text{ilr}(\mathbf{X}_i) = \frac{1}{n} \sum_{i=1}^n \mathbf{X}_i^*.$$

When $\mathbf{X}_1, \mathbf{X}_2, \ldots, \mathbf{X}_n, \ldots$ are independently, identically, normally distributed, that is, $\mathbf{X}_i \sim \mathcal{N}_S(\boldsymbol{\mu}, \boldsymbol{\Sigma})$ for any i, Theorems 6.20 and 6.21 apply and the average is also normal, that is, $\overline{\mathbf{X}}_n \sim \mathcal{N}_S(\boldsymbol{\mu}, (1/n)\boldsymbol{\Sigma})$ that mimics the well-known result for the real multivariate variables. This will be later used in Section 7.1 to formally establish estimators of these parameters.

Theorem 6.20 can be generalized to general linear transformations (endomorphisms, Section 4.9.2) of a normal random compositions. It can be stated as follows.

Theorem 6.23. *Let* $\mathbf{X} \sim \mathcal{N}_S(\boldsymbol{\mu}, \boldsymbol{\Sigma})$ *be a random composition in* S^D *in which covariance matrix* $\boldsymbol{\Sigma}$ *has been represented with respect to an orthogonal basis defined by a contrast matrix* $\boldsymbol{\Psi}$. *Consider an endomorphism of* S^D, *represented as the* (D, D)-*matrix* $\boldsymbol{B}$ *such that the elements of its columns and its rows add to zero, and a constant composition* $\mathbf{a} \in S^D$. *Let* Y *be the random composition in* S^D *obtained as the affine transformation of* $\mathbf{X}$: $\mathbf{Y} = \mathbf{a} \oplus (\mathbf{X} \boxdot \boldsymbol{B})$. *Then,* $\mathbf{Y} \sim \mathcal{N}_S(\boldsymbol{\mu}_Y, \boldsymbol{\Sigma}_Y)$, *with* $\mathrm{cen}[\mathbf{Y}] = \mathbf{a} \oplus \mathrm{cen}[\mathbf{X}] \boxdot \boldsymbol{B}$ *and parameters*

$$\boldsymbol{\mu}_Y = \mathrm{ilr}(\mathbf{a}) + \boldsymbol{\mu} \cdot \boldsymbol{B}^*, \quad \boldsymbol{\Sigma}_Y = (\boldsymbol{B}^*)^\mathsf{T} \boldsymbol{\Sigma} \boldsymbol{B}^*,$$

with $\boldsymbol{B}^* = \boldsymbol{\Psi} \boldsymbol{B} \boldsymbol{\Psi}^\mathsf{T}$.

The most important property of the multivariate normal distribution is the *central limit theorem*. In a formulation for ilr-coordinates, it states how the n-average $\overline{\mathbf{X}}_n^*$ of n random vectors has a normal limiting distribution. Consider a sequence of n independent real random vectors in $\mathbb{R}^{D-1}$, $\mathbf{X}_k^*$, $k = 1, 2, \ldots, n, \ldots$, with $\mathrm{E}[\mathbf{X}_k^*] = \boldsymbol{\mu}$ and $\mathrm{cov}[\mathbf{X}_k^*] = \boldsymbol{\Sigma}$. Then,

$$\mathbf{Z}_n^* = \sqrt{n}(\overline{\mathbf{X}}_n^* - \boldsymbol{\mu}),$$

where $\overline{\mathbf{X}}_n^*$ is the n-average of the sequence, has a multivariate distribution converging in distribution, as $n \to \infty$, to $\mathcal{N}(\mathbf{0}, \boldsymbol{\Sigma})$ (Kocherlakota and Kocherlakota, 2004). Convergence in distribution or law is a weak convergence. Roughly speaking, for each value of the real vector $\mathbf{x}^*$, the cdf of the variable $\mathbf{Z}_n^*$ converges to the limiting distribution in $\mathbf{x}^*$. This means that the large n necessary for a good approach of the cdf can change significantly from a value of $\mathbf{x}^*$ to another. Typically, for a fixed n, the approximation of the cdf is better in the central part of the distribution, that is, near $\boldsymbol{\mu}$. This was a good reason for calling the result *central limit theorem*. The conditions on this statement can be relaxed (e.g., Chow and Teicher, 1997; Rao, 1973; Feller, 1966) but the simple formulation given here is enough for the present aim.

The central limit theorem for coordinates of a random composition can be translated into a compositional formulation (Aitchison, 1986; Mateu-Figueras et al., 2013). The ilr-transformation is assumed fixed, so that the covariance matrix $\boldsymbol{\Sigma}$ is the covariance matrix of the ilr-coordinates denoted by $\mathbf{X}_i^*$; the center of a random composition is $\boldsymbol{\eta} = \mathrm{ilr}^{-1}(\boldsymbol{\mu})$.

Theorem 6.24   (Central limit theorem in the simplex).
Let $\mathbf{X}_i$, $i = 1, 2, \ldots, n, \ldots$ be a sequence of independent random compositions with common center and covariance matrix, $\mathrm{cen}[\mathbf{X}_i] = \boldsymbol{\eta}$ and $\mathrm{cov}[\mathbf{X}_i^*] = \boldsymbol{\Sigma}$. Consider the n average in the simplex $\overline{\mathbf{Z}}_n = (1/n) \odot \bigoplus_{i=1}^{n} \mathbf{X}_i$. The distribution of the random composition

$$\sqrt{n} \odot (\overline{\mathbf{Z}}_n \ominus \boldsymbol{\eta})$$

converges in law, as $n \to \infty$, to the distribution $\mathcal{N}_S(\mathbf{0}, \boldsymbol{\Sigma})$.

The consequence of the central limit theorem is that a multivariate normal distribution is likely to be a good approximation of a sum of real, independent, identically distributed random variables. The same occurs for random compositions, but the operation *sum* is changed to *perturbation*. The following fictitious, but realistic, example shows that situations in which the normal distribution on the simplex appears can be frequent.

A random composition with a normal distribution in the simplex can be obtained as the closure of D jointly lognormal random variables. This fact was described in Aitchison (1986). Here it is presented using ilr techniques that illustrate the use of orthogonal coordinates of the simplex. Consider the vector of D random variables $\mathbf{X} = [X_1, X_2, \ldots, X_D]$, supported in $\mathbb{R}_+^D$, with joint lognormal pdf (with respect to the Lebesgue measure)

$$f_{\mathbf{X}}(\mathbf{x}) = \frac{1}{\left(\prod_{i=1}^{D} x_i\right)(2\pi)^{D/2}|\mathbf{C}|^{-1/2}}$$
$$\cdot \exp\left(-\frac{1}{2}[\ln(\mathbf{x}) - \boldsymbol{\mu}]\mathbf{C}^{-1}[\ln(\mathbf{x}) - \boldsymbol{\mu}]^\top\right) \cdot \mathrm{I}\{\mathbf{x} \in \mathbb{R}_+^D\}, \tag{6.7}$$

where $\boldsymbol{\mu}$ is the logarithmic multivariate mean and $\mathbf{C}$ is the logarithmic covariance matrix assumed nonsingular, that is, its determinant $|\mathbf{C}| \neq 0$. The random composition $C\mathbf{X}$ can be represented in coordinates of the simplex S^D. For instance, the sequential binary partition (SBP) signs code shown in Table 6.1 can be used for the ilr transformation, and let $\boldsymbol{\Psi}$ denote the $(D-1, D)$ contrast matrix associated with this SBP. For an easy transformation of the joint pdf (Equation 6.7) of D random variables, an auxiliary coordinate x_D^* is required. To this end, define

$$x_D^* = \frac{1}{\sqrt{D}} \sum_{i=1}^{D} \ln x_i = \ln \prod_{i=1}^{D-1} x_i^{\sqrt{D}},$$

Table 6.1   Signs code of the sequential binary partition used in the balance-coordinate representation of the joint lognormal pdf

	x_1	x_2	x_3	...	x_{D-1}	x_D
x_1^*	+1	−1	−1	...	−1	−1
x_2^*	0	+1	−1	...	−1	−1
...	...	...	...	...	...	...
x_{D-1}^*	0	0	0	...	+1	−1

which allows the definition of an $\mathbb{R}^D \to \mathbb{R}^D$ mapping $x_D^* = \ln(x)\ \boldsymbol{\Psi}_D^{\mathsf{T}}$, where $x_D^* = [x^* \mid x_D^*]$ is the enlarged vector of coordinates and the bordered (D, D)-matrix is

$$\boldsymbol{\Psi}_D = \begin{pmatrix} & & \boldsymbol{\Psi} & \\ ... & ... & ... & ... \\ 1/\sqrt{D} & 1/\sqrt{D} & ... & 1/\sqrt{D} \end{pmatrix}, \quad \boldsymbol{\Psi}_D^{\mathsf{T}}\boldsymbol{\Psi}_D = \mathbf{I}_D, \quad |\boldsymbol{\Psi}_D| = 1,$$

that is, $\boldsymbol{\Psi}_D$ is an orthogonal matrix and $\boldsymbol{\Psi}_D^{\mathsf{T}}$ is its inverse (Egozcue et al., 2013). The joint distribution of the new variables in $\mathbf{X}_D^*$ is obtained by substituting $\ln(x)$ as a function of x_D^* in Equation (6.7) and taking into account the Jacobian of the transformation $x_D^* = \ln(x)\ \boldsymbol{\Psi}_D^{\mathsf{T}}$. The value of the Jacobian is proven to be

$$\left|\frac{\partial x_D^*}{\partial x}\right| = \frac{1}{\prod_{i=1}^D x_i} \cdot \left|\boldsymbol{\Psi}_D^{\mathsf{T}}\right| = \frac{1}{\prod_{i=1}^D x_i}.$$

Taking into account that $\boldsymbol{\Psi}_D^{\mathsf{T}}\boldsymbol{\Psi}_D = \mathbf{I}_D$ and that x_D^* span the whole $\mathbb{R}^D$, the pdf of the random variables $\mathbf{X}_D^*$ is then

$$f_{\mathbf{X}_D^*}(x_D^*) = \frac{(\prod_{i=1}^D x_i)}{(\prod_{i=1}^D x_i)(2\pi)^{D/2}|\mathbf{C}|^{-1/2}}$$
$$\cdot \exp\left(-\frac{1}{2}[\ln(x) - \mu]\boldsymbol{\Psi}_D^{\mathsf{T}}(\boldsymbol{\Psi}_D \mathbf{C}^{-1}\boldsymbol{\Psi}_D^{\mathsf{T}})\boldsymbol{\Psi}_D[\ln(x) - \mu]^{\mathsf{T}}\right)$$
$$= \frac{1}{(2\pi)^{D/2}|\mathbf{C}|^{-1/2}}$$
$$\cdot \exp\left(-\frac{1}{2}[x_D^* - \mu\boldsymbol{\Psi}_D^{\mathsf{T}}](\boldsymbol{\Psi}_D \mathbf{C}^{-1}\boldsymbol{\Psi}_D^{\mathsf{T}})[x_D^* - \mu\boldsymbol{\Psi}_D^{\mathsf{T}}]^{\mathsf{T}}\right), \tag{6.8}$$

that is, the extended random vector of coordinates $\mathbf{X}_D^*$ has a multivariate normal distribution in $\mathbb{R}^D$, with mean $\boldsymbol{\mu}\boldsymbol{\Psi}_D^\mathsf{T}$ and covariance matrix $\boldsymbol{\Psi}_D\mathbf{C}\boldsymbol{\Psi}_D^\mathsf{T}$. Note that the inverse of this covariance matrix is $\boldsymbol{\Psi}_D\mathbf{C}^{-1}\boldsymbol{\Psi}_D^\mathsf{T}$ and that $|\mathbf{C}| = |\boldsymbol{\Psi}_D\mathbf{C}\boldsymbol{\Psi}_D^\mathsf{T}|$. The random vector of coordinates in the simplex, $\mathbf{X}^*$, is also distributed normal because its distribution is a marginal of the joint pdf in Equation (6.8), which is in turn jointly normal.

6.3.2   The Dirichlet distribution

The Dirichlet distribution (Aitchison, 1986, p.58) is obtained as the closure of independent, equally scaled, gamma-distributed, positive random variables. To construct the density, D-independent, gamma-distributed random variables Z_i, $i = 1, 2, \ldots, D$ are considered. Their marginal densities are

$$f_i(z_i|\alpha_i, \nu) = \frac{\nu^{\alpha_i}}{\Gamma(\alpha_i)} z_i^{\alpha_i-1} \ \exp(-\nu z_i) \cdot \mathrm{I}\{z_i > 0\}, \ \ i = 1, 2, \ldots, D,$$

where $\nu > 0$ is a scale parameter of the gamma distribution, common to all Z_i; $\boldsymbol{\alpha} = (\alpha_1, \alpha_2, \ldots, \alpha_D)$ are positive shape parameters of the gamma distributions; and the indicator function $\mathrm{I}\{z_i > 0\}$ is 0 for $z_i \leq 0$ and 1 for $z_i > 0$, thus pointing out that the random variables Z_i are positive. Interest can be centered on the proportions, $X_i = Z_i/(\sum_k Z_k), i = 1, 2, \ldots, D$, and their joint distribution. Note that X_i cannot be independent as they add up to 1. The joint distribution of $\mathbf{X} = [X_1, X_2, \ldots, X_D]$, known as the *Dirichlet distribution*, corresponds to the density

$$f_L(\mathbf{x}|\alpha) = \frac{\Gamma(\alpha_1 + \alpha_2 + \ldots + \alpha_D)}{\Gamma(\alpha_1) \cdot \Gamma(\alpha_2) \ldots \Gamma(\alpha_D)} x_1^{\alpha_1-1} x_2^{\alpha_2-1} \ldots x_D^{\alpha_D-1} \cdot \mathrm{I}\{\mathbf{x} \in S^D\}, \qquad (6.9)$$

where $\Gamma(\cdot)$ is the Euler gamma function, and the indicator function $\mathrm{I}\{\mathbf{x} \in S^D\}$ is 1 whenever $\mathbf{x} \in S^D$ and 0 elsewhere. Note that the scale parameter ν does not appear in the Dirichlet density in Equation (6.9), as it cancels out in the proportions X_i. The Dirichlet density, $f_L(\mathbf{x})$ (Equation 6.9), is expressed as a density with respect to the Lebesgue measure in $\mathbb{R}^D$, fact that is pointed out by the subscript L.

The first interesting property of the Dirichlet distribution is that it is strictly supported on the simplex S^D and its density (Equation 6.9) is easily expressed as a function of the proportions. As $\mathbf{x} \in S^D$, one of the proportions, say x_D, can be substituted by $x_D = 1 - \sum_{i=1}^{D-1} x_i$, as frequently presented in textbooks. When $D = 2$, the Dirichlet distribution reduces to the beta distribution on the interval $[0, 1]$ or, equivalently, on S^2.

A second relevant family of properties are subcompositional invariance and amalgamation invariance. Consider $Sub(\cdot)$ an application that selects some indices of a composition. Thus, a random subcomposition $\mathbf{X}_S$ (Definition 2.6) is built from a random composition $\mathbf{X}$ through $\mathbf{X}_S = CSub(\mathbf{X})$. Then, if $\mathbf{X}$ has the Dirichlet distribution with parameter $\boldsymbol{\alpha}$, the distribution of a random subcomposition $\mathbf{X}_S$ is again of the Dirichlet family, with parameter vector $Sub(\boldsymbol{\alpha})$. Interestingly, the same happens with amalgamation. If $Am(\cdot)$ denotes an application that forms an amalgamated composition (Definition 2.6), and $\mathbf{X}$ follows the Dirichlet distribution, then $Am(\mathbf{X})$ follows as well the Dirichlet distribution with parameter $Am(\boldsymbol{\alpha})$.

Analogously to the normal in the simplex, the Dirichlet distribution can be described by a density with respect to the Aitchison measure in S^D, $f_a(\mathbf{x}) = f_L(\mathbf{x})J(\mathbf{x})$, with $J(\mathbf{x}) = \sqrt{D} \cdot \prod_i x_i$ (Monti et al., 2011). Denoting $\alpha_+ = \sum_{i=1}^{D} \alpha_i$, this density is

$$f_a(\mathbf{x}|\boldsymbol{\alpha}) = \frac{\sqrt{D}\Gamma(\alpha_+)}{\Gamma(\alpha_1) \cdot \Gamma(\alpha_2) \ldots \Gamma(\alpha_D)} x_1^{\alpha_1} x_2^{\alpha_2} \ldots x_D^{\alpha_D} \cdot \mathrm{I}\{\mathbf{x} \in S^D\}. \tag{6.10}$$

Figure 6.2 shows the contours of the two densities $f_L(\mathbf{x}|\boldsymbol{\alpha})$ and $f_a(\mathbf{x}|\boldsymbol{\alpha})$ for the case $D = 3$ and $\boldsymbol{\alpha} = (1.3, 1.7, 2.0)$. Important differences are visualized. These differences correspond to the change of location and dispersion parameters. For instance, the expectation and mode with respect to the Lebesgue measure are

$$\mathrm{E}_L[\mathbf{X}] = \left[\frac{\alpha_1}{\alpha_+}, \frac{\alpha_2}{\alpha_+}, \ldots, \frac{\alpha_D}{\alpha_+} \right], \tag{6.11}$$

$$\mathrm{Mode}_L[\mathbf{X}] = \left[\frac{\alpha_1 - 1}{\alpha_+ - D}, \frac{\alpha_2 - 1}{\alpha_+ - D}, \ldots, \frac{\alpha_D - 1}{\alpha_+ - D} \right], \tag{6.12}$$

which are not equal to those obtained using the Aitchison geometry:

$$\mathrm{E}_a[\mathbf{X}] = C[\exp[\psi(\alpha_1)], \exp[\psi(\alpha_2)], \ldots, \exp[\psi(\alpha_D)]], \tag{6.13}$$

$$\mathrm{Mode}_a[\mathbf{X}] = \left[\frac{\alpha_1}{\alpha_+}, \frac{\alpha_2}{\alpha_+}, \ldots, \frac{\alpha_D}{\alpha_+} \right], \tag{6.14}$$

where ψ denotes the Euler digamma function, the logarithmic derivative of the Euler gamma function. The expectation $\mathrm{E}_a[\mathbf{X}]$ is the center of the random composition X, that is, $\mathrm{E}_a[\mathbf{X}] = \mathrm{cen}[\mathbf{X}]$. These location parameters are also shown in Figure 6.2. It should be noted that $\mathrm{E}_L[\mathbf{X}]$, although meaningless as a mean when the Aitchison geometry is recognized as natural in the simplex, coincides with the mode in that geometry.

The same happens with the variance of the Dirichlet distribution. Considering them as real variables with respect to the Lebesgue measure, the variances and covariances are

$$\mathrm{var}[X_i] = \frac{\alpha_i(\alpha_+ - \alpha_i)}{\alpha_+^2(\alpha_+ + 1)}, \quad \mathrm{cov}[X_i, X_j] = \frac{-\alpha_i \alpha_j}{\alpha_+^2(\alpha_+ + 1)}.$$

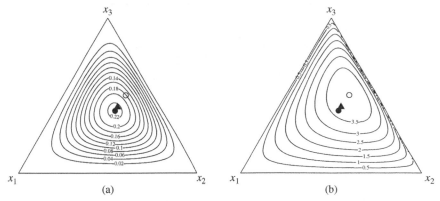

Figure 6.2   Contours of probability density functions of the Dirichlet random composition with parameters α = (1.3, 1.7, 2.0). Filled triangle, center. Filled circle, mode in Aitchison geometry and also mean with respect to the Lebesgue measure. Empty circle, mode with respect to the Lebesgue measure. (a) pdf with respect to the Aitchison measure. (b) pdf with respect to the Lebesgue measure.

Table 6.2   Signs code of the sequential binary partition used in the balance-coordinate representation of the Dirichlet pdf in Figure 6.3

	x_1	x_2	x_3
x_1^*	−1	+1	0
x_2^*	−1	−1	+1

Given a contrast matrix $\boldsymbol{\Psi}$, the covariance matrix of the associated coordinates of X is $\boldsymbol{\Sigma} = \boldsymbol{\Psi}\mathrm{diag}[\psi'(\boldsymbol{\alpha})]\boldsymbol{\Psi}^\top$, where $\psi'(\cdot)$ is the Euler trigamma function, applied componentwise to the vector of parameters $\boldsymbol{\alpha}$, and $\mathrm{diag}[\cdot]$ is a diagonal matrix (Boogaart and Tolosana-Delgado, 2013).

The Dirichlet pdf can be represented in coordinates. The signs code of the SBP used is shown in Table 6.2. Two features are visible. The pdf contours are far from having an elliptical shape, and the mode of the distribution with respect to the Lebesgue measure in the simplex is clearly deviated from the mode and expectation with respect to the Aitchison measure.

The Dirichlet distribution has been extensively used in Bayesian statistics, as it is the conjugate distribution of the multinomial likelihood (see Section 7.5). Furthermore, the Dirichlet distribution has the apparent advantage of being supported in the simplex in its simplest form, shown in Equation (6.9), in such a

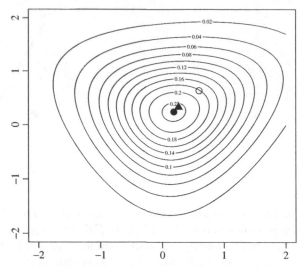

Figure 6.3   Contours in coordinates of the Dirichlet probability density function with parameters $\alpha = (1.3, 1.7, 2.0)$. *Filled triangle, center. Filled circle, mode in the Aitchison geometry and also mean with respect to the Lebesgue measure. Empty circle, mode with respect to the Lebesgue measure.*

way that no transformation is needed. These two circumstances made the Dirichlet distribution quite popular (see references at the beginning of Section 6.3). However, it is a quite inflexible model, as can be deduced from its genesis by closing a set of independent, gamma-distributed, random variables, with equal scale parameter. In comparison to the construction of the logistic-normal distribution (Section 6.3.1), the independence of the gamma-distributed random variables seems to be the main reason for the lack of flexibility. This lack of adaptability to data makes it difficult to find compositional data sets fitting the Dirichlet distribution. The independence of the gamma-distributed parts generating the Dirichlet distribution points out a kind of maximal independence of the Dirichlet parts, as extensively discussed in Aitchison (1986) [Chapter 13].

Example 6.25. In the synthesis of a couple of valuable liquid organic compounds (labeled B and C), another nuisance compound of the same liquid nature (named A) is also produced. After their synthesis, it is necessary to concentrate the valuable ones for further processing and putting on sale. For this goal, a complex filter is used, comprising several slices of the same provider. In the product specifications, the filter provider states that the slice filters let pass 5% of component A, as well as 20% of B and 30% of C. It is also guaranteed that 95% of the effluent contains less than 20% of A. The action of a slice of filter can be assumed to be a random perturbation $\mathbf{F} = [F_A, F_B, F_C]$, so that the effluent composition is $\mathbf{y} = \mathbf{F} \oplus \mathbf{x}$, where $\mathbf{x}$ is the affluent composition, with an average filter perturbation $\hat{\mathbf{F}} = C[5, 20, 30]$.

The question is which is the retentive effect $\mathbf{G}$ (transfer function) of the filter made of $n = 25$ filter pieces? The transfer function $\mathbf{G}$ is characterized by the probability distribution of $\mathbf{G}$ in the simplex S^3 and its center and covariance matrix in some ilr coordinates. Taking into account that $\mathbf{G}$ is the perturbation of n random perturbations $\mathbf{F}_i$, $i = 1, 2, \ldots, n$, which can be assumed to be independent and equally distributed, the perturbation $\mathbf{G}$ will converge to a normal distribution on the simplex as a consequence of the central limit theorem. Its parameters (mean and covariance matrix) for a given orthonormal basis can be estimated from the mean and covariance matrix of the $\mathbf{F}_i$ perturbations.

The first step is to assess the distribution of $\mathbf{F}_i$. The provider does not state interactions between the filtering capacity corresponding to different parts; consequently, the Dirichlet distribution can be a model for the random transfer function $\mathbf{F}_i$. On one hand, it is said that the average output is proportional to $\hat{\mathbf{F}} = C[5, 20, 30]$. Assuming that the provider has never heard about compositional statistics, it can be considered that these numbers are arithmetic averages, which would imply that $\hat{\mathbf{F}}$ approximates $F_L[\mathbf{X}]$. This implies that $\alpha_2/\alpha_1 = 20/5$ and $\alpha_3/\alpha_1 = 30/5$. This is satisfied by $[A, B, C] = 5 : 20 : 30$, and also by $[A, B, C] = 1 : 4 : 6$, and by many other combinations of three numbers in these relative proportions, scaled by a constant value α_+. To find out which scaling α_+ has to be taken, consider that in 95% of the slice filters, the retained

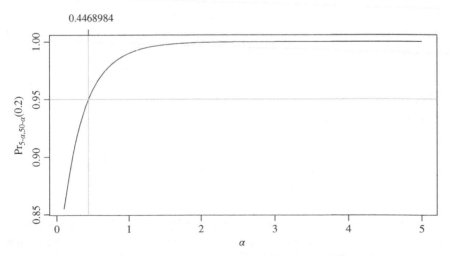

Figure 6.4   *Finding the scaling constant α_+ of Example 6.25 on the axis labeled α.*

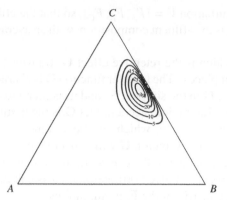

Figure 6.5   Density contours of the Dirichlet distribution of the transfer function F_i for each slide filter with respect to the Lebesgue measure.

mass contains less than 20% of A. Owing to the marginalizing properties of the Dirichlet distribution, it is known that the distribution of F_A is a beta distribution with parameters 5α and $(20 + 30)\alpha$ such that $P[F_A < 0.2] = 0.95$. The probability contained within the interval $(0, 0.2)$ of a beta-distributed random variable with parameters 5α and 50α can be computed with the cumulative probability function (available in Excel or OpenOffice as function BETADIST, or in R as function pbeta). Figure 6.4 graphically shows that the sought scaling $\alpha_+ = 0.44$, which means that each individual filter F_i follows the Dirichlet distribution with parameters $[2.23, 8.94, 13.41]$. Figure 6.5 shows the distribution of the effluent filtering values of one single filter of this kind. Assuming independence and equal distribution of the transfer functions of each slice, the complete filter made of n pieces has transfer function $G = F_1 \oplus F_2 \oplus \ldots F_n = n \cdot \overline{F}$. As a random composition, G has a distribution that can be approximated by a normal on the simplex because of the central limit theorem (Theorem 6.24). The centers of $\overline{F}$ and G are the same as the center of each $F_i = \text{clr}^{-1}[\psi(2.23), \psi(8.94), \psi(13.41)]$, while the variance matrix of $\overline{F}$ must be the variance of F_i

$$\text{var}[F_i] = \Psi \cdot \begin{pmatrix} \psi'(2.23) & 0 & 0 \\ 0 & \psi'(8.94) & 0 \\ 0 & 0 & \psi'(13.41) \end{pmatrix} \cdot \Psi^{\mathsf{T}}$$

in any particular ilr basis scaled by a factor $1/\sqrt{n}$. Therefore, the variance of G must be the same scaled by $n/\sqrt{n} = \sqrt{n}$. Figure 6.6 shows the approximate contours of the pdf of G for $n = 25$, both in a ternary diagram and in the ilr coordinate space, built applying the assumption of normality.   ◇

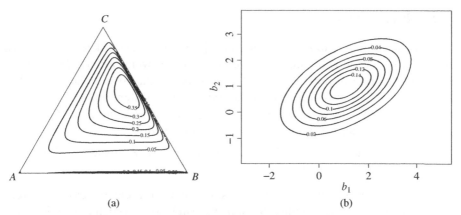

(a)   (b)

*Figure 6.6   Density contours of the distribution of **G**, the transfer function of a complex n = 25 slice filter effluent. (a) With respect to the Aitchison measure in the ternary diagram. (b) Density contours of the ilr-coordinates. Note that this transfer function is not the average of n = 25 slices, but their perturbation, as it has more variability than an individual slice (Figure 6.5).*

6.3.3   Other distributions

Many other distributions on the simplex have been defined as modifications of the Dirichlet distribution. For instance, the multivariate beta distribution or mixtures of Dirichlet distributions (Mosimann, 1962; Connor and Mosimann, 1969; Dishon and Weiss, 1980). A related distribution is the Aitchison distribution (Aitchison, 1986; Tolosana-Delgado et al., 2011), that is obtained as a posterior distribution in a multinomial sampling, when the prior is a logistic normal (a normal on the simplex). Further modifications of the Dirichlet distribution have been obtained by perturbation (also called scaled-Dirichlet) and powering of a Dirichlet random composition (Monti et al., 2011).

A rich way to introduce distributions in the simplex is that used in Section 6.3.1: consider a distribution in $\mathbb{R}^{D-1}$, for example, the normal distribution, the multivariate t-distribution, or the skew-normal distribution, as the distribution of the coordinates of a composition; the backtransformation into the simplex provides a distribution for compositions. This way of construction was initiated in Aitchison and Shen (1980) and Aitchison (1982) for the logistic-normal distribution inspired in previous lognormal techniques (Aitchison and Brown, 1957). A general study of normal distributions in different sample spaces can be found in (Eaton, 1983). Using similar techniques, Mateu-Figueras and Pawlowsky-Glahn (2007) introduced the skew-normal distribution on the simplex and Mateu-Figueras et al. (2013) reviewed the technique for normal distributions on $\mathbb{R}_+$ and on the simplex S^D paying attention to the change of reference measure.

6.4 Exercises

Exercise 45. [*Sample space of ipsative data*] An individual is asked to order three items, x_1, x_2, x_3, following his/her preference. The answer is arranged in an array in which the order of preference, 1, 2, 3, is placed in the position labeled x_i. For instance, $[3, 1, 2,]$ means that x_2 is the most preferred item and x_1 is the third one preferred. Nonresponse is excluded. The random variable $\mathbf{Z} = [Z_1, Z_2, Z_3]$, containing the orders of preference, is then considered.

(a) Determine the minimal sample space, $\mathcal{M}$, for $\mathbf{Z}$, made of all possible triplets $\mathbf{z}_1, \mathbf{z}_2, \ldots, \mathbf{z}_m$.

(b) The analyst decided that the possible values of $\mathbf{Z}$ can be reasonably represented in the simplex $\mathcal{S}^3$ and the Aitchison geometry is valid. Give reasons supporting these ideas.

(c) Accepting statement (b), compute the Aitchison interdistances between the compositions $\mathbf{z}_1, \mathbf{z}_2, \ldots, \mathbf{z}_m$.

(d) The probabilities of getting the answer $\mathbf{z}_i$ are given. Consider, for instance, two cases. The first is $p_i = 1/m$, $i = 1, 2, \ldots, m$. The second is

$$q_i = \binom{m-1}{i-1} q^{i-1}(1-q)^{m-i}, \quad q = 0.9, \quad i = 1, 2, \ldots, m.$$

Compute the center and total variance of $\mathbf{Z}$, when the sample space is $\mathcal{M}$. Repeat the computation when $\mathcal{M}$ is assumed embedded in $\mathcal{S}^3$.

Exercise 46. Plot the density of a normal on $\mathcal{S}^2$. Do it for several logistic means and variances. For instance, $\mu = 0$, $\mu = -1$, and logistic variances $\sigma^2 = 0.5^2$, $\sigma^2 = 2^2$. These densities can be plotted taking the Lebesgue measure as reference measure in $[0, 1]$ and, alternatively, taking the Aitchison measure on $\mathcal{S}^2$ identified with $[0, 1]$.

Exercise 47. Prove Theorem 6.10.

Exercise 48. Prove Theorem 6.19.

Exercise 49. Using Theorem 6.23, find a way to transform a standard normal distribution (i.e., with ilr mean vector $\mu = \mathbf{0}$ and ilr variance $\Sigma = \mathbf{I}_{D-1}$) to another normal distribution on the simplex with arbitrary ilr mean and variance. Assume that this arbitrary variance matrix is of the form $\Sigma = \mathbf{L} \cdot \mathbf{L}^\mathsf{T}$.

Exercise 50. Using the expression obtained in Exercise 49, simulate 100 compositions following a normal distribution on the simplex with geometric center and variation matrix

$$
\mathbf{g} = \begin{pmatrix} 0.089 \\ 0.239 \\ 0.672 \end{pmatrix}, \quad \mathbf{T} = \begin{pmatrix} 0.000 & 0.405 & 2.938 \\ 0.405 & 0.000 & 2.911 \\ 2.938 & 2.911 & 0.000 \end{pmatrix},
$$

Note that you should first transform the statistics given to a mean vector and variance matrix of a particular ilr of your choice (see Section 5.5). Note as well that the matrix $\mathbf{L}$ can be obtained with the Cholesky decomposition, after transforming the variation matrix to an ilr-variance matrix. (see Sections A.2 and A.3).

Exercise 51. Given the definition of the Dirichlet distribution of Section 6.3.2, it seems natural to simulate the Dirichlet composition as the closure of D gamma-distributed values, with the same v and each with its corresponding α_i. Using this procedure, for $D = 3$, simulate $N = 100$ realizations of the Dirichlet-distributed composition with $\boldsymbol{\alpha} = (1.3, 1.7, 2.0)$. Note that v can be any arbitrary number. Plot the resulting simulations in a ternary diagram. Compare the results with Figure 6.2.

7

Statistical inference

7.1 Point estimation of center and variability

In Chapters 5 and 6, center and variability of a sample and a random composition were defined. In particular, the normal distribution on the simplex was characterized by the mean vector and covariance matrix of coordinates with respect to a given basis of the simplex (Section 6.3.1). It is thus pertinent to consider which is the relationship between these sample statistics and distributional parameters and, in particular, how the latter can be estimated. To cut a long story short, the center parameter is estimated with the sample center, the coordinate mean parameter with the sample ilr mean, and the covariance matrix parameter with the sample ilr covariance. The theory of parameter estimation is mainly based on the concept of likelihood function both in frequentist and Bayesian statistics (see, e.g., Jeffreys, 1961; Box and Tiao, 1973; Robert, 1994; Shao, 1998). The estimation of center and variability of a random composition can also be developed from this point of view.

In general, if a random variable $\mathbf{X}$ follows a known probability model that depends on a vector of unknown parameters θ and a random sample $\{\mathbf{x}_1, \mathbf{x}_2, \ldots, \mathbf{x}_n\}$ is available, the likelihood function of the parameters is computed as

$$L(\theta | \mathbf{x}_1, \mathbf{x}_2, \ldots, \mathbf{x}_n) = \prod_{k=1}^{n} f_{\mathbf{X}}(\mathbf{x}_k | \theta), \tag{7.1}$$

Modeling and Analysis of Compositional Data, First Edition.
Vera Pawlowsky-Glahn, Juan José Egozcue and Raimon Tolosana-Delgado.
© 2015 John Wiley & Sons, Ltd. Published 2015 by John Wiley & Sons, Ltd.

being $f_{\mathbf{X}}(\mathbf{x}|\theta)$ the probability density function (pdf) of $\mathbf{X}$ given θ. The likelihood function (7.1) corresponds to a sample in which the elements of the sample are independent once the value of the parameter θ is given. Otherwise, the likelihood function is the joint pdf of the sample given the observations.

Two cases are of particular importance for the analysis of compositions: the estimation of the parameters of a normal distribution on the simplex and a multinomial distribution. In both cases, the estimation of parameters using the maximum likelihood (ML) method is illustrative. The ML estimator of θ is a function $\hat{\theta}_{ML}$ of the available sample, for which the likelihood function is maximized, that is,

$$L(\hat{\theta}_{ML}|\mathbf{x}_1, \mathbf{x}_2, \ldots, \mathbf{x}_n) = \sup_{\theta}\{L(\theta|\mathbf{x}_1, \mathbf{x}_2, \ldots, \mathbf{x}_n)\},$$

where θ spans the space of parameters when looking for the supreme. ML estimators have remarkable asymptotic properties, which make them the preferred family of estimators. However, ML estimation requires a model for the random variable. Also ML estimators can lose their general properties when, for instance, the supreme value is not a regular point of the likelihood function (e.g., Rohatgi, 1976). In most cases, the function that is maximized is the logarithm of the likelihood function, or log-likelihood, due to the fact that transforming a function by a strictly monotone function maintains the extreme points.

The two examples, the normal distribution on the simplex and the multinomial distribution, have different relationship with the simplex. In the case of the normal distribution on the simplex, the sample space is the simplex and the parameters are a real vector (mean coordinates) and a positive definite matrix (covariance matrix). In the case of the multinomial distribution, the observations, which are counts, span a sample space of nonnegative vectors of integers, but the parameters constitute a composition.

Example 7.1 Assume that the random composition $\mathbf{X}$ follows a normal distribution on the simplex as specified in Definition 6.18. The vector of parameters θ contains the coordinate mean vector $\boldsymbol{\mu}$ and the coordinate covariance matrix $\boldsymbol{\Sigma}$. Given a sample of size n, their joint likelihood is

$$L(\boldsymbol{\mu}, \boldsymbol{\Sigma}|\mathbf{x}_1, \ldots, \mathbf{x}_n) = \prod_{k=1}^{n} f_{\mathbf{X}}(\mathbf{x}_k|\theta) = \left(\frac{1}{(2\pi)^{(D-1)/2} \ |\boldsymbol{\Sigma}|^{1/2}}\right)^n$$

$$\times \prod_{k=1}^{n} \exp\left[-\frac{1}{2}(\mathrm{ilr}(\mathbf{x}_k) - \boldsymbol{\mu})\boldsymbol{\Sigma}^{-1}(\mathrm{ilr}(\mathbf{x}_k) - \boldsymbol{\mu})^{\top}\right], \quad (7.2)$$

or, taking logs and removing constant additive terms,

$$l(\boldsymbol{\mu}, \boldsymbol{\Sigma}|\mathbf{x}_1, \ldots, \mathbf{x}_n) = -\frac{n}{2} \cdot \ln |\boldsymbol{\Sigma}| - \frac{1}{2}\sum_{k=1}^{n}(\mathrm{ilr}(\mathbf{x}_k) - \boldsymbol{\mu})\boldsymbol{\Sigma}^{-1}(\mathrm{ilr}(\mathbf{x}_k) - \boldsymbol{\mu})^{\top}. \quad (7.3)$$

Taking derivatives in Equation (7.3) with respect to μ and Σ and equating the result to zero, one obtains the estimators described in Theorems 7.2 and 7.3. Proofs of these standard results of multivariate normal inference can be found in Krzanowski (1988), for instance. $\diamond$

Theorem 7.2   (Maximum likelihood estimator of center).
Let $\mathbf{X}$ *be a random normal composition with mean parameter* μ. *Let* $\{\mathbf{x}_1, \mathbf{x}_2, \ldots, \mathbf{x}_n\}$ *be a random sample of* $\mathbf{X}$. *Then, the ML estimators of the center* cen[$\mathbf{X}$] *(Definition 6.9) and the mean parameter* μ *are*

$$\widehat{\text{cen}}[\mathbf{X}] = \hat{\mathbf{g}} = \mathcal{C}\left(\prod_{k=1}^{n} \mathbf{x}_k\right)^{1/n} = \frac{1}{n} \odot \bigoplus_{k=1}^{n} \mathbf{x}_k,$$

$$\hat{\mu} = \overline{\mathbf{x}}^* = \frac{1}{n} \sum_{k=1}^{n} \text{ilr}(\mathbf{x}_k),$$

where $\hat{\mathbf{g}}$ *and* $\overline{\mathbf{x}}^*$ *are, respectively, the sample center (Definition 5.1) and the sample* ilr *mean (Section 5.5).*

Recall that both center and mean coordinate vector are related through expressions given by Equation (5.4) for the sample version and in Theorem 6.10 for the parameters, so that estimating the center implies estimating the mean ilr-coordinate vector and vice versa. The same applies to the mean vector of the clr-transformed compositions.

Theorem 7.3   (Maximum likelihood estimation of variability).
Let $\mathbf{X}$ *be a random composition with covariance matrix parameter* Σ *specified in an* ilr *coordinate system given by the contrast matrix* $\mathbf{\Psi}$. *Let* $\{\mathbf{x}_1, \mathbf{x}_2, \ldots, \mathbf{x}_n\}$ *be a sample of* $\mathbf{X}$. *The ML estimator of the covariance matrix* Σ *is*

$$\hat{\Sigma} = \hat{\mathbf{S}},$$

where $\hat{\mathbf{S}}$ *is the sample* ilr-*coordinate covariance matrix*

$$\hat{\mathbf{S}} = \frac{1}{n} \sum_{k=1}^{n} (\text{ilr}(\mathbf{x}_k) - \text{ilr}(\hat{\mathbf{g}}))^{\mathsf{T}} \cdot (\text{ilr}(\mathbf{x}_k) - \text{ilr}(\hat{\mathbf{g}})).$$

The estimator of Σ considered in Theorem 7.3 has alternative expressions corresponding to different representations of the random composition,

$$\hat{\Sigma} = \hat{\mathbf{S}} = \mathbf{\Psi}\hat{\mathbf{S}}^c\mathbf{\Psi}^{\mathsf{T}} = -\frac{1}{2}\mathbf{\Psi}\hat{\mathbf{T}}\mathbf{\Psi}^{\mathsf{T}},$$

where $\hat{\mathbf{T}} = [\hat{t}_{ij}]$ is the sample variation matrix,

$$\hat{t}_{ij} = \frac{1}{n} \sum_{k=1}^{n} \left(\ln \frac{x_{ki}}{x_{kj}} - \ln \frac{\hat{g}_i}{\hat{g}_j} \right)^2,$$

and $\hat{\mathbf{S}}^c = [s_{ij}^c]$ is the sample clr covariance matrix,

$$\hat{\mathbf{S}}^c = \frac{1}{n} \sum_{k=1}^{n} (\mathrm{clr}(\mathbf{x}_k) - \mathrm{clr}(\hat{\mathbf{g}}))^\top \cdot (\mathrm{clr}(\mathbf{x}_k) - \mathrm{clr}(\hat{\mathbf{g}})).$$

These alternative expressions are sample versions of the properties that relate the variability representations of a random compositions (See Section A.3).

Example 7.4 Let $\mathbf{Z} = [Z_1, Z_2, \cdots, Z_D]$ be a random vector of counts on D categories, and assume that $\mathbf{Z}$ follows a multinomial distribution with compositional parameter $\mathbf{p}$ in S^D and total number of trials $N = Z_1 + Z_2 + \cdots + Z_D$. The composition $\mathbf{p}$ contains the probabilities p_i of counting one event in the ith category in one multinomial trial. The probability function of the multinomial distribution is

$$f_\mathbf{Z}(\mathbf{z}|\mathbf{p}) = P[Z_1 = z_1, Z_2 = z_2, \ldots, Z_D = z_D] = N! \prod_{i-1}^{D} \frac{p_i^{z_i}}{z_i!},$$

for $z_1 + z_2 + \cdots + z_D = N$.

When observing the results of N multinomial trials, the obtained sample is the number of times each category was obtained, arranged in the vector $\mathbf{z} = [z_1, z_2, \ldots, z_D]$, satisfying $z_1 + z_2 + \cdots + z_D = N$. Note that $\mathbf{z}$ is a sample of size equal to 1. The likelihood, $L(\mathbf{p}|\mathbf{z})$ is $f_\mathbf{Z}(\mathbf{z}|\mathbf{p})$ when $\mathbf{z}$ is the observed sample and it is considered as a function of $\mathbf{p}$, that is, it is a function such that $L(\cdot|\mathbf{z}) : S^D \rightarrow [0, 1]$. The log-likelihood of $\mathbf{p}$ is

$$l(\mathbf{p}|\mathbf{z}_1, \ldots, \mathbf{z}_n) = \ln N! - \sum_{i=1}^{D} \ln z_i! + \sum_{i=1}^{D} z_i \ln p_i, \qquad (7.4)$$

where only the last term is a function of $\mathbf{p}$; the remaining terms can be removed for optimization. Note that the vectors of counts $\mathbf{z}_j$ is not a composition in the sense of Definition 2.1, because its components are natural numbers, possibly including zeroes. When a vector of nonnegative components contains one or more zeroes, the relative information involving null parts is lost, as some ratios become infinite or null. However, the parameter $\mathbf{p}$ is in the simplex S^D, as it has no zeroes among its components. To maximize Equation (7.4), the constrain that the sum of p_i add to 1 must be taken into account. There are different ways of carrying out the optimization of the log-likelihood (Equation 7.4). An elegant way is to constrain the maximization to $\sum_i p_i = 1$ using a Lagrange multiplier. This consists of maximization of the function

$$H(\mathbf{p}) = \sum_{i=1}^{D} z_i \ln\, p_i + \lambda\left(1 - \sum_{i=1}^{D} p_i \right),$$

where the first term is the log-likelihood function (7.4) after removing constant terms and λ is a Lagrange multiplier associated with the condition that the probabilities add to 1. Taking derivatives of $H(\mathbf{p})$ with respect to p_i, $i = 1, 2, \ldots, D$ and equating to zero for optimization, it yields

$$\frac{z_i}{p_i} - \lambda = 0, \quad i = 1, 2, \ldots, D.$$

The value of the Lagrange multiplier is $\lambda = N$ as derived when summing up all these equations. Then, each p_i can be isolated. The resulting ML estimator is then $\hat{p}_i = z_i/N$, that is, the sample relative frequency of each category. The estimated composition is

$$\hat{\mathbf{p}}_{ML} = \left[\frac{z_1}{N}, \frac{z_2}{N}, \ldots, \frac{z_D}{N} \right]. \tag{7.5}$$

This is the ML estimator of a compositional parameter, given a sample of multinomial counts. Unfortunately, specially when D is large and the total number of counts N is small, this procedure can yield estimates of some p_i equal to zero. This result is normally unacceptable, as the multinomial model assumes that all categories are possible and, accordingly, with positive probability. The null estimation of a probability can be problematic as examined in Section 2.3. This unsatisfactory result is revisited in Section 7.5, where an alternative Bayesian estimation of the parameter vector $\mathbf{p}$ is introduced. ◇

The properties of estimators of compositional parameters (unbiasedness, consistency, sufficiency, etc.) may be directly established by making use of the principle of working on coordinates: the properties of coordinate parameter estimators will be inherited by the compositional estimators. For brevity, only definition of bias and root-mean-square error of an estimator of a compositional parameter are given, following the development in Pawlowsky-Glahn and Egozcue (2001, 2002).

Definition 7.5 (Simplicial bias).
The (simplicial) bias of an estimator $\hat{\theta}$ of a compositional parameter θ is

$$\mathrm{sBias}[\hat{\theta}] = \mathrm{cen}[\mathrm{cen}[\hat{\theta}] \ominus \theta].$$

Definition 7.6 (Simplicial root-mean-square error).
The (simplicial) root-mean-square error of an estimator $\hat{\theta}$ of a compositional parameter θ is

$$\text{sRMSE}[\hat{\theta}] = \sqrt{\text{E}[\text{d}_a^2(\hat{\theta}, \theta)]}.$$

Property 7.7

1. The center $\hat{\mathbf{g}}$ as an estimator of $\text{ilr}^{-1}(\boldsymbol{\mu})$:

   (a) is (simplicially) unbiased in the simplex, that is, $\text{sBias}[\hat{\mathbf{g}}] = \mathbf{n}$, and

   (b) has the smallest (simplicial) root-mean-square error possible among all possible estimators of central tendency.

2. The empirical ilr covariance matrix, $\hat{\mathbf{S}}$, is not a composition and has the same properties as any covariance estimator of a conventional multivariate framework. In particular, ilr/clr-covariance matrices and variation matrices estimated with denominator n are ML estimators in the normal case, but they are biased, while those estimated with denominator $(n - 1)$ are unbiased.

3. The ML estimator $\hat{\mathbf{p}}$ (Equation 7.5) of a multinomial distribution is (simplicially) biased, that is, $\text{sBias}[\hat{\mathbf{g}}] \neq \mathbf{n}$. The unbiased estimator is actually of the form of Equation (6.13), as will be shown in Section 7.5.

4. The center and one of the variance estimators mentioned earlier are sufficient statistics of a normal distribution on the simplex, and the ML estimator $\hat{\mathbf{p}}$ is a sufficient statistic of a multinomial population (Shao, 1998).

7.2   Testing hypotheses on compositional normality

The most widely used statistical methods and models assume all a normal distribution, univariate, or multivariate. All compositional models presented in this book are not an exception. Thus, it is necessary to have a way to check for evidences that a compositional data set does or does not follow a normal distribution on the simplex (Section 6.3.1). Following the principle of working on coordinates, this could be done testing that a set of ilr coordinates follows a joint multivariate normal, with any classical conventional goodness-of-fit test. For instance, the energy test of Székely and Rizzo (2005) can be readily used to check that any ilr-transformed composition follows a joint $(D - 1)$-dimensional distribution, which would imply that the composition itself would follow a normal distribution on the simplex. Unfortunately, such multivariate normality tests are not widely known and only available in advanced statistical software e.g., in R (Rizzo and Székely, 2008). For these reasons, this section presents another approach, based on the singular value decomposition (SVD).

Consider a compositional data set $\mathbf{X} = [x_{ij}]$ and its clr-transformed, centered version $\mathbf{Z} = [z_{ij}]$. Following Section 5.4.1, an SVD of this last matrix provides a set of $(D - 1)$ normalized scores $\mathbf{U} = [u_{ik}]$ for each observation. They are standardized ilr-coordinates of the sample. Accordingly, if $\mathbf{X}$ follows a normal distribution on the simplex, then $\mathbf{U} = \mathbf{Z} \cdot (\mathbf{V} \cdot \mathbf{D}^{-1})$ follows a $(D - 1)$-variate normal distribution with standard, uncorrelated components. This can be tested along three levels (Aitchison et al., 2004):

1. *Univariate tests of marginal normality.* The data on each column of $\mathbf{U}$ might be tested for standard normality, with any available normality test (Kolmogorov–Smirnoff, Anderson–Darling, etc.).

2. *Bivariate tests of uncorrelation.* In the absence of correlation, each pair of columns of $\mathbf{U}$ defines a vector pointing randomly (uniformly) on any direction. Thus, the random angle $\tan^{-1}(u_{ij}/u_{ik})$ defined between columns j and k can be tested for fit with a uniform distribution (with Kolmogorov–Smirnoff tests, for instance).

3. *Radius tests.* Finally, the square norm of the rows of $\mathbf{U}$, $(u_{i1}^2 + u_{i2}^2 + \cdots + u_{i(D-1)}^2)^{1/2}$, being the sum of $(D - 1)$ squared standard normal variables, should follow a χ^2 distribution with $(D - 1)$ degrees of freedom. This can also be tested with a Kolmogorov–Smirnoff test.

Note that the significance of these tests should be adjusted for joint testing: $D - 1$ for the marginal tests, $(D - 1)(D - 2)/2$ for the angular tests, and 1 for the global radius tests give a total of $D(D - 1)/2 - 1$ tests. In these cases of multiple testing, the significance level should be adjusted following standard techniques (e.g., Holm, 1979).

A final comment is due with regard to this procedure of compositional normality testing: in principle, one should know the mean and the variability of the compositional data to apply them, as their statistical significance is only accurate if the distributional parameters are known. Given this fact and the uncertainty around the multiple testing correction, it seems just reasonable to take these tests as indicators of gross departure from compositional normality.

7.3   Testing hypotheses about two populations

When a sample has been divided into two or more populations, interest may lie in finding out whether there is a difference between those populations and, if it is the case, whether it is due to differences in the center, the covariance structure, or both. Consider for simplicity two samples of size n_1 and n_2, which are realization of two random compositions $\mathbf{X}_1$ and $\mathbf{X}_2$, each with a normal distribution on the simplex, a distributional assumption that should have been assessed before. Let $\boldsymbol{\mu}_i$, $\boldsymbol{\Sigma}_i$, $i = 1, 2$, be the mean and covariance matrix of the coordinates of the respective random compositions. The comparison of the two populations can be performed by stating the following hypotheses:

(a) there is no distributional difference between both populations;

(b) the covariance structure is the same, but the centers are different;

(c) the centers are the same, but the covariance structure is different;

(g) the populations differ in their centers and covariance structure,

where the label (g) stands for general hypothesis. Note that if the first hypothesis (a) is not rejected, it makes no sense to consider (b) or (c); the same happens for the hypothesis (b) and (c) with respect to hypothesis (g). This can be considered as a lattice structure in which one goes from lowest level of complexity (a), to the highest level (g), until he or she is unable to reject one of the hypotheses. At that point, it makes no sense to test further hypothesis and it is advisable to stop.

To perform tests on these hypotheses, sample coordinates $\mathbf{x}^*$ are used, and it is assumed that each of them follows a multivariate normal distribution. For the parameters of the two multivariate normal distributions, the three null hypotheses and the general hypothesis are expressed, in the same order as above, as follows:

$$\mathcal{H}_a: \boldsymbol{\mu}_1 = \boldsymbol{\mu}_2 \text{ and } \boldsymbol{\Sigma}_1 = \boldsymbol{\Sigma}_2;$$

$$\mathcal{H}_b: \boldsymbol{\mu}_1 \neq \boldsymbol{\mu}_2 \text{ and } \boldsymbol{\Sigma}_1 = \boldsymbol{\Sigma}_2;$$

$$\mathcal{H}_c: \boldsymbol{\mu}_1 = \boldsymbol{\mu}_2 \text{ and } \boldsymbol{\Sigma}_1 \neq \boldsymbol{\Sigma}_2;$$

$$\mathcal{H}_g: \boldsymbol{\mu}_1 \neq \boldsymbol{\mu}_2 \text{ and } \boldsymbol{\Sigma}_1 \neq \boldsymbol{\Sigma}_2.$$

The last hypothesis is called the general model and taken as alternative hypothesis; the other hypotheses will be tested against it.

Under multivariate normality assumptions, ML estimates of the parameters can be used to state test statistics. Also unbiased estimators of the covariance matrix can be used as, for sample size n, divisions by n are changed to $n - 1$ in the unbiased case. Here developments are presented in terms of ML estimates, as those have been used in the previous chapters. Note that estimators change under each of the possible hypotheses and, accordingly, each case will be presented separately. The following developments are based on Aitchison (1986, pp. 153–158) and Krzanowski (1988, pp. 323–329), although for a complete theoretical proof, Mardia et al. (1979, Section 5.5.3) is recommended. To proceed with tests, the

n_i vectors of coordinates, ilr($\mathbf{x}_i$) = $\mathbf{x}_i^*$, $i = 1, 2$ from the two populations are computed. From these sample coordinates, the statistics to be computed are

- the separate population estimates of

  1. the vectors of mean values:

$$\hat{\boldsymbol{\mu}}_1 = \frac{1}{n_1} \sum_{r=1}^{n_1} \mathbf{x}_{1r}^*, \quad \hat{\boldsymbol{\mu}}_2 = \frac{1}{n_2} \sum_{s=1}^{n_2} \mathbf{x}_{2s}^*,$$

  2. the covariance matrices:

$$\hat{\boldsymbol{\Sigma}}_1 = \frac{1}{n_1} \sum_{r=1}^{n_1} (\mathbf{x}_{1r}^* - \hat{\boldsymbol{\mu}}_1)^{\mathsf{T}} (\mathbf{x}_{1r}^* - \hat{\boldsymbol{\mu}}_1),$$

$$\hat{\boldsymbol{\Sigma}}_2 = \frac{1}{n_2} \sum_{s=1}^{n_2} (\mathbf{x}_{2s}^* - \hat{\boldsymbol{\mu}}_2)^{\mathsf{T}} (\mathbf{x}_{2s}^* - \hat{\boldsymbol{\mu}}_2),$$

- the pooled covariance matrix estimate:

$$\hat{\boldsymbol{\Sigma}}_p = \frac{n_1 \hat{\boldsymbol{\Sigma}}_1 + n_2 \hat{\boldsymbol{\Sigma}}_2}{n_1 + n_2},$$

- the combined sample estimates:

$$\hat{\boldsymbol{\mu}}_c = \frac{n_1 \hat{\boldsymbol{\mu}}_1 + n_2 \hat{\boldsymbol{\mu}}_2}{n_1 + n_2},$$

$$\hat{\boldsymbol{\Sigma}}_c = \hat{\boldsymbol{\Sigma}}_p + \frac{n_1 n_2 (\hat{\boldsymbol{\mu}}_1 - \hat{\boldsymbol{\mu}}_2)^{\mathsf{T}} (\hat{\boldsymbol{\mu}}_1 - \hat{\boldsymbol{\mu}}_2)}{(n_1 + n_2)^2}.$$

To test the different hypotheses, the generalized likelihood ratio test will be used; it is based on the following principles: consider the maximized likelihood function for data $\mathbf{x}^*$ under the null hypothesis, $L_0(\mathbf{x}^*)$ and under the model with no restrictions (general model), $L_g(\mathbf{x}^*)$. The test statistic is then $R(\mathbf{x}^*) = L_g(\mathbf{x}^*)/L_0(\mathbf{x}^*)$, and the larger the value is, the more critical or resistant not to reject the null hypothesis one should be. In some cases, the exact distribution $R(\mathbf{x}^*)$ is known. In cases where it is not known, Wilks' asymptotic approximation can be used: under the null hypothesis, which places c constraints on the parameters, the test statistic $Q(\mathbf{x}^*) = 2 \ln (R(\mathbf{x}^*))$ is distributed approximately as $\chi^2(c)$ (Rohatgi, 1976). For the cases to be studied, the approximate generalized ratio test statistic takes the form

$$Q(\mathbf{x}^*) = n_1 \ln \left(\frac{|\hat{\boldsymbol{\Sigma}}_{10}|}{|\hat{\boldsymbol{\Sigma}}_{1g}|} \right) + n_2 \ln \left(\frac{|\hat{\boldsymbol{\Sigma}}_{20}|}{|\hat{\boldsymbol{\Sigma}}_{2g}|} \right). \tag{7.6}$$

Table 7.1 Hypotheses of equality of means and covariances, with their numbers of free parameters and restrictions with respect to the general hypothesis. For compositions of D parts, $v_\mu = D - 1$ and $v_\Sigma = D(D-1)/2$. Estimators are indicated with the symbol $\simeq$. Subscripts p and c stand for pooled and combined estimators; h points out that the estimator is obtained using iterative methods.

	Means	Covariances	Parameters	Restrictions (c)
$\mathcal{H}_a$	$\mu_1 = \mu_2 \simeq \hat{\mu}_c$	$\Sigma_1 = \Sigma_2 \simeq \hat{\Sigma}_c$	$v_\mu + v_\Sigma$	$v_\mu + v_\Sigma = \frac{(D-1)(D+2)}{2}$
$\mathcal{H}_b$	$\mu_1 \simeq \hat{\mu}_1$	$\Sigma_1 = \Sigma_2 \simeq \hat{\Sigma}_p$	$2v_\mu + v_\Sigma$	$v_\Sigma = \frac{D(D-1)}{2}$
	$\mu_2 \simeq \hat{\mu}_2$			
$\mathcal{H}_c$	$\mu_1 = \mu_2 \simeq \hat{\mu}_h$	$\Sigma_1 \simeq \hat{\Sigma}_{1h}$	$v_\mu + 2v_\Sigma$	$v_\mu = D - 1$
		$\Sigma_2 \simeq \hat{\Sigma}_{2h}$		
$\mathcal{H}_g$	$\mu_1 \simeq \hat{\mu}_1$	$\Sigma_1 \simeq \hat{\Sigma}_1$	$2v_\mu + 2v_\Sigma$	0
	$\mu_2 \simeq \hat{\mu}_2$	$\Sigma_2 \simeq \hat{\Sigma}_2$		

A summary of these statistics and the number of restrictions of each hypothesis are given in Table 7.1 and described in detail in the following points:

(a) *Equality of centers and covariance structure.* The null hypothesis is that $\mu_1 = \mu_2$ and $\Sigma_1 = \Sigma_2$, thus estimates of the common parameters $\mu = \mu_1 = \mu_2$ and $\Sigma = \Sigma_1 = \Sigma_2$ are needed. They are, respectively, $\hat{\mu}_c$ for μ and $\hat{\Sigma}_c$ for Σ under the null hypothesis and $\hat{\mu}_i$ for μ_i and $\hat{\Sigma}_i$ for Σ_i, $i = 1, 2$, under the model, resulting in a test statistic

$$Q_{\text{a.vs.g}}(\mathbf{x}^*) = n_1 \ln \left(\frac{|\hat{\Sigma}_c|}{|\hat{\Sigma}_1|} \right) + n_2 \ln \left(\frac{|\hat{\Sigma}_c|}{|\hat{\Sigma}_2|} \right) \sim \chi^2 \left(\frac{1}{2}(D-1)(D+2) \right).$$

The upper tail probability of $Q_{\text{a.vs.g}}(\mathbf{x}^*)$ as a quantile is the p-value of the test.

(b) *Equality of covariance structure with different centers.* The null hypothesis is that $\mu_1 \neq \mu_2$ and $\Sigma_1 = \Sigma_2$, thus the estimates of μ_1, μ_2 and the common covariance matrix $\Sigma = \Sigma_1 = \Sigma_2$ are needed; they are $\hat{\Sigma}_p$ for Σ under the null hypothesis and $\hat{\Sigma}_i$ for Σ_i, $i = 1, 2$, under the model, resulting in a test statistic

$$Q_{\text{b.vs.g}}(\mathbf{x}^*) = n_1 \ln \left(\frac{|\hat{\Sigma}_p|}{|\hat{\Sigma}_1|} \right) + n_2 \ln \left(\frac{|\hat{\Sigma}_p|}{|\hat{\Sigma}_2|} \right) \sim \chi^2 \left(\frac{1}{2}D(D-1) \right).$$

(c) *Equality of centers with different covariance structure.* The null hypothesis is that $\mu_1 = \mu_2$ and $\Sigma_1 \neq \Sigma_2$, thus the estimates of the common center

$\mu = \mu_1 = \mu_2$ and the covariance matrices Σ_1 and Σ_2 are needed. In this case, no explicit form for the ML estimates is available. A simple iterative method provides a solution. It requires the following steps:

1. Set the initial value, for $h = 0$, $\hat{\Sigma}_{ih} = \hat{\Sigma}_{i0} = \hat{\Sigma}_i, i = 1, 2$;

2. compute the common mean, weighted by the variance of each population:

$$\hat{\mu}_h = (n_1 \hat{\Sigma}_{1h}^{-1} + n_2 \hat{\Sigma}_{2h}^{-1})^{-1} (n_1 \hat{\Sigma}_{1h}^{-1} \hat{\mu}_1 + n_2 \hat{\Sigma}_{2h}^{-1} \hat{\mu}_2);$$

3. compute the variances of each group with respect to the common mean:

$$\hat{\Sigma}_{ih} = \hat{\Sigma}_i + (\hat{\mu}_i - \hat{\mu}_h)^\top (\hat{\mu}_i - \hat{\mu}_h), i = 1, 2;$$

4. Repeat steps 2 and 3 until convergence.

This results in $\hat{\Sigma}_{i0}$ for $\Sigma_i, i = 1, 2$, under the null hypothesis and $\hat{\Sigma}_i$ for Σ_i, $i = 1, 2$, under the model, leading to a test statistic

$$Q_{\text{c.vs.g}}(\mathbf{x}^*) = n_1 \ln \left(\frac{|\hat{\Sigma}_{1h}|}{|\hat{\Sigma}_1|} \right) + n_2 \ln \left(\frac{|\hat{\Sigma}_{2h}|}{|\hat{\Sigma}_2|} \right) \sim \chi^2(D - 1).$$

Example 7.8 Section 5.8 presented an exploratory data analysis of nutrition patterns in European countries (data in Table 5.5). The question is now to test whether the northern and southern block countries have significant differences. The two groups have, respectively, $n_N = 16$ and $n_S = 9$ countries. For the sake of simplicity, two three-part subcompositions will be considered: (starch, eggs, milk) and (nuts, fish, white meat). For each of these subcompositions, the ilr basis is used, which is defined by the contrast matrix

$$\Psi = \begin{pmatrix} -\frac{1}{\sqrt{2}} & +\frac{1}{\sqrt{2}} & 0 \\ -\frac{1}{\sqrt{6}} & -\frac{1}{\sqrt{6}} & +\frac{2}{\sqrt{6}} \end{pmatrix}.$$

For the subcomposition (starch, eggs, milk), group means and variances are computed,

$$\hat{\mu}_N = [-0.246, 1.286] \quad \hat{\mu}_S = [-0.279, 1.316],$$

$$\hat{\Sigma}_N = \begin{pmatrix} 0.0580 & 0.0156 \\ 0.0156 & 0.0930 \end{pmatrix} \quad \hat{\Sigma}_S = \begin{pmatrix} 0.2272 & 0.1536 \\ 0.1536 & 0.2737 \end{pmatrix},$$

while the combined variance, computed pooling together all samples, is

$$\hat{\Sigma}_c = \begin{pmatrix} 0.1123 & 0.0607 \\ 0.0607 & 0.14961 \end{pmatrix}.$$

The statistic $\hat{Q}_{a.vs.g} = 5.2129$ is compared with a $\chi^2(5)$ with $5 = (3-1) \cdot (3 + 2)/2$ degrees of freedom, giving a p-value of 0.3904. This p-value is larger than the typical critical value of 0.05, then hypothesis $\mathcal{H}_a$ of equality of means and variances is not rejected: southern and northern countries do not exhibit a significantly different pattern in the subcomposition (starch, eggs, milk).

Repeating the calculations for the subcomposition (nuts, fish, white meat) leads to the group estimators

$$\hat{\mu}_N = [0.472, 1.089] \qquad \hat{\mu}_S = [-0.529, 0.214],$$

$$\hat{\Sigma}_N = \begin{pmatrix} 0.7775 & -0.1761 \\ -0.1761 & 0.2907 \end{pmatrix} \qquad \hat{\Sigma}_S = \begin{pmatrix} 1.0750 & -0.3073 \\ -0.3073 & 0.442 \end{pmatrix},$$

with the combined and pooled variances

$$\hat{\Sigma}_c = \begin{pmatrix} 1.0847 & -0.0023 \\ -0.0023 & 0.5129 \end{pmatrix}, \qquad \hat{\Sigma}_p = \begin{pmatrix} 0.8846 & -0.2233 \\ -0.2233 & 0.345 \end{pmatrix},$$

The hypothesis of equality of mean and variance yields a statistic $Q_{a.vs.g} = 20.174$, which according to a $\chi^2(5)$ reference distribution gives a p-value of 0.00116, small enough to reject $\mathcal{H}_a$. On the contrary, the hypothesis of equal variance but different mean has a p-value of 0.866, derived from a statistic $Q_{b.vs.g} = 0.732$, to be compared with a $\chi^2(3)$ with $0.5 \cdot 3 \cdot (3-1) = 3$ degrees of freedom. In this case, there is no reason to reject $\mathcal{H}_b$.

In summary, southern and northern European countries center consumption in the subcomposition (starch, eggs, milk) are statistically similar, while within the subcomposition (nuts, fish, white meat), there is a significant different center consumption between northern and southern countries. Both groups of countries show a similar variance structure, with no significant differences. ◇

Example 7.9 (A structural zero example).
Following the political example in Section 5.9, consider the effect of the appearance of party C's in the distribution of votes to the other parties. Remember that before 2005, five main parties contested in the parliament elections, while that year, C's appeared as a new political force and apparently changed the political landscape of this region. The question is whether that change is significant statistically. To avoid considering also trends in time, the focus is placed on 2 years elections: 2003 and 2006.

Two groups or populations are considered: districts where C's did not get any vote and districts where this party got at least one vote. The numbers of observations for each of these two groups are 41 in both the cases (the number of electoral districts). Using the coordinates proposed in Section 5.9, group means and variances give

$$\hat{\mu}_{2003} = [-0.335 - 1.057 - 1.2220.321] \, ,$$

$$\hat{\Sigma}_{2003} = \begin{pmatrix} 0.091 & 0.047 & 0.022 & -0.093 \\ 0.047 & 0.072 & 0.001 & -0.059 \\ 0.022 & 0.001 & 0.050 & 0.001 \\ -0.093 & -0.059 & 0.001 & 0.127 \end{pmatrix} \, ,$$

$$\hat{\mu}_{2006} = [-0.350 - 1.111 - 0.8370.158] \, ,$$

$$\hat{\Sigma}_{2006} = \begin{pmatrix} 0.065 & 0.037 & -0.002 & -0.067 \\ 0.037 & 0.082 & -0.015 & -0.061 \\ -0.002 & -0.015 & 0.046 & 0.012 \\ -0.067 & -0.061 & 0.012 & 0.122 \end{pmatrix} \, ,$$

while the combined variance is

$$\hat{\Sigma}_c = \begin{pmatrix} 0.077 & 0.042 & 0.008 & -0.078 \\ 0.042 & 0.077 & -0.012 & -0.057 \\ 0.008 & -0.012 & 0.085 & -0.009 \\ -0.078 & -0.057 & -0.009 & 0.130 \end{pmatrix} \, ,$$

These values give a statistic $Q_{a.vs.g} = 87.111$, for the hypothesis of equal means and variances, associated with a p-value of $1.329208 \cdot 10^{-12}$ under a $\chi^2(14)$ reference distribution. Consequently, $\mathcal{H}_a$ is rejected. But the hypothesis $\mathcal{H}_b$ of equal variances and different means cannot be rejected, as the test statistic $Q_{b.vs.g} = 15.298$, linked to a p-value of 0.121 for a $\chi^2(10)$. In summary, the appearance of C's changed the relations between the other parties in mean (center) but there is no evidence that any change occurred in their variability structure.   ◇

7.4   Probability and confidence regions for normal data

Like confidence intervals for univariate variables, confidence regions are a measure of variability, although in the case of compositional data, it is a measure of joint variability for several variables. They can be of interest in themselves, to analyze the precision of the estimation obtained, but more frequently, they are used to visualize differences between groups or populations. Recall that for compositional data with three components, confidence regions can be plotted in the corresponding ternary diagram, thus giving evidence of the relative behavior of the various centers or the populations themselves. Alternatively, confidence regions can be plotted in a plane of two ilr coordinates, thus obtaining a proper representation of distances. The following method to compute confidence regions assumes either multivariate normality of coordinates or the size of the sample to be large enough for the multivariate central limit theorem to hold.

Consider a composition $\mathbf{x} \in S^D$ and assume it follows a normal distribution on S^D as defined in Section 6.18. Then, the $(D-1)$-variate vector $\mathbf{X}^* = \text{ilr}(\mathbf{X})$ follows a multivariate normal distribution. As will be seen in Chapter 8, this can be either observable data or the estimated parameters of a given linear model.

In any case, three different cases are of interest:

1. the mean vector and the variance matrix of the random vector $\mathbf{X}^*$ are known, and the goal is to plot a probability region for the possible values of $\mathbf{X}^*$;

2. the mean vector and variance matrix of the $\mathbf{X}^*$ are not known, and the goal is to plot a confidence region for its mean using a sample of size n;

3. the mean vector and variance matrix of $\mathbf{X}^*$ are not known, and the goal is to plot a probability region for $\mathbf{X}^*$, incorporating all uncertainty coming from the estimation of the mean and variance.

In the first case, if a random vector $\mathbf{X}^*$ follows a $(D-1)$-multivariate normal distribution with known parameters $\boldsymbol{\mu}$ and $\boldsymbol{\Sigma}$, then

$$(\mathbf{x}^* - \boldsymbol{\mu})\boldsymbol{\Sigma}^{-1}(\mathbf{x}^* - \boldsymbol{\mu})^\top \sim \chi^2(D-1)$$

is a χ^2 distribution of $D-1$ degrees of freedom. Thus, for a given α, a value κ can be obtained (through software or tables) such that

$$1 - \alpha = P[(\mathbf{x}^* - \boldsymbol{\mu})\boldsymbol{\Sigma}^{-1}(\mathbf{x}^* - \boldsymbol{\mu})^\top \leq \kappa]. \tag{7.7}$$

This defines a $(1-\alpha)100\%$ probability region centered at $\boldsymbol{\mu}$ in $\mathbb{R}^D$ and, consequently, $\mathbf{x} = \text{ilr}^{-1}(\mathbf{x}^*)$ defines a $(1-\alpha)100\%$ probability region centered at the mean in the simplex.

Regarding the second case, it is well known that for $\mathbf{X}^*$ normally distributed or n large enough, the ML estimates of the mean $\boldsymbol{\mu}$, $\overline{\mathbf{x}}^*$, and $\hat{\boldsymbol{\Sigma}}$ satisfy that

$$\frac{n-D+1}{D-1}(\overline{\mathbf{x}}^* - \boldsymbol{\mu})\hat{\boldsymbol{\Sigma}}^{-1}(\overline{\mathbf{x}}^* - \boldsymbol{\mu})^\top \sim \mathcal{F}(D-1, n-D+1)$$

follows a Fisher $\mathcal{F}$ distribution on $(D-1, n-D+1)$ degrees of freedom (Krzanowski, 1988, see pp. 227–228 for further details). Again, for given α, we may obtain a value c such that

$$1 - \alpha = P\left[\frac{n-D+1}{D-1}(\overline{\mathbf{x}}^* - \boldsymbol{\mu})\hat{\boldsymbol{\Sigma}}^{-1}(\overline{\mathbf{x}}^* - \boldsymbol{\mu})^\top \leq c\right]$$

$$= P[(\overline{\mathbf{x}}^* - \boldsymbol{\mu})\hat{\boldsymbol{\Sigma}}^{-1}(\overline{\mathbf{x}}^* - \boldsymbol{\mu})^\top \leq \kappa], \tag{7.8}$$

with $\kappa = c(D-1)/(n-D+1)$. But $(\overline{\mathbf{x}}^* - \boldsymbol{\mu})\hat{\boldsymbol{\Sigma}}^{-1}(\overline{\mathbf{x}}^* - \boldsymbol{\mu})^\top = \kappa$ (constant) defines a $(1-\alpha)100\%$ confidence region centered at $\overline{\mathbf{x}}^*$ in $\mathbb{R}^D$. This region can be transformed back to S^D and cen$[\mathbf{X}] = \text{ilr}^{-1}(\boldsymbol{\mu})$ describes a $(1 - \alpha)100\%$ confidence region around the center in the simplex.

Finally, in the third case, the predictive distribution is a case of the multivariate student distribution (Aitchison, 1986, p. 174). The probability density of a new realization of $\mathbf{X}^*$ has density

$$f(\mathbf{x}^*|data) \propto \left[1 + \frac{n}{n^2 - 1}(\mathbf{x}^* - \overline{\mathbf{x}}^*)\hat{\boldsymbol{\Sigma}}^{-1}(\mathbf{x}^* - \overline{\mathbf{x}}^*)^\top\right]^{-n/2}.$$

This distribution is not commonly tabulated, and it is only available in some specific packages. On the other side, if n is large with respect to D, the differences between the first and third studied cases are negligible.

Note that for $D = 3, D - 1 = 2$ and we have an ellipse in real space, in any of the three cases: the only difference between them is how the constant κ is computed. The parameterization equations in polar coordinates, which are necessary to plot these ellipses, are given in Appendix B.

7.5   Bayesian estimation with count data

Most of this book follows the *frequentist* approach to statistics, where probability is thought to be an *objective* measure of the frequency of occurrence of each possible outcome of a random phenomenon. In this view, if an event has not occurred yet, its probability of occurrence is estimated as zero. This framework may not be adequate in some circumstances, for instance, when estimating the parameter vector of a multinomial distribution with a small number of counts (as mentioned in Example 7.4) or in problems that require latent variables (Section 2.3.2). The alternative is to work within the framework of Bayesian statistics (Jeffreys, 1961; Robert 1994), where the probability of an event is considered as a *subjective* assessment of its chances of occurrence. Consequently, this probability of occurrence is modeled as a random variable.

In Bayesian statistics, no fundamental differences are made between data $\mathbf{X}$ (observable results of any experiment) and model parameters θ describing the nature of the system under study: both are considered random variables. Before any observation is obtained, one takes all information available about the possible values of the parameters gathered into a *prior distribution* $f_{\boldsymbol{\Theta}}^0(\theta)$. One then plans an experiment, which result is a random variable $\mathbf{X}$ following a known model given the parameters, that is, one knows the probability density function $f_{\mathbf{X}}(\mathbf{x}|\theta)$ for all values of θ. Conducting the experiment, one obtains a certain result $\mathbf{x}$, which allows to compute the *posterior distribution* of the parameters given the data as

$$f_{\boldsymbol{\Theta}}(\theta|\mathbf{x}) = C \cdot f_{\boldsymbol{\Theta}}^0(\theta) \cdot f_{\mathbf{X}}(\mathbf{x}|\theta), \tag{7.9}$$

where C is a normalizing constant that makes $f_\Theta(\theta|\mathbf{x})$ integrate to 1. Equation (7.9) is the Bayes theorem for absolutely continuous random observations and parameters (see also Theorem B.8). One of the advantages of the Bayesian approach is that one finally obtains a whole description of the possible values of θ, with its uncertainty and most likely values. An equivalent to a point estimate $\hat{\theta}$ can be obtained by extracting some central tendency statistic of $f_\Theta(\theta|\mathbf{x})$, for example, the mode (maximum posterior estimate) or its mean value (posterior expectation). But one can also draw regions of probability following the posterior distribution to show the uncertainty of the parameters, as in Section 7.4.

Modern computer-based Bayesian statistics does not force any shape or condition on the possible prior for θ or experimental model for $\mathbf{X}$. But, traditionally, it was customary to choose models for θ and $\mathbf{X}$ in such a way that the form of the posterior $f_\Theta(\theta|\mathbf{x})$ was known analytically. These are called *conjugate priors*, and we have already introduced three cases: multinomial-Dirichlet, multinomial-Aitchison (mentioned in Section 6.3.3), and normal-Student/Siegel (mentioned in Section 7.4). The following is an example of the first case.

After Example 7.4, consider the random vector $\mathbf{X} = [X_1, X_2, \cdots, X_D]$ containing the number of times that each of the D categories have been found out of N multinomial experiments. The random vector of counts $\mathbf{X}$ is assumed to follow a multinomial distribution with compositional parameter $\mathbf{p}$ and $N = X_1 + X_2 + \ldots + X_D$, that is, with density

$$f_\mathbf{X}(\mathbf{x}|\mathbf{p}) = N! \prod_{i=1}^{D} \frac{p_i^{x_i}}{x_i!}, \quad \sum_{i=1}^{D} x_i = N.$$

Assume a prior distribution for $\mathbf{p}$ of the Dirichlet family (Equation 6.9), which is characterized with its own parameter vector $\boldsymbol{\alpha}^0$. If one obtains a multinomial sample $\mathbf{x}$ containing counts of N trials, it can be shown that the posterior distribution for $\mathbf{p}$ is also of the Dirichlet family, with posterior parameter $\boldsymbol{\alpha}$,

$$\boldsymbol{\alpha} = \boldsymbol{\alpha}^0 + \mathbf{x}.$$

A point estimate $\hat{\theta}$ can be derived with any of the expressions for means and modes of the Dirichlet distribution given in Equations (6.11)–(6.14), depending on the choice of the analyst. Regions of uncertainty can be drawn then from the Dirichlet density as in Figure 6.2.

The fact that the parameter of the multinomial distribution is $\mathbf{p} \in S^D$ suggests to express it using ilr-coordinates corresponding to some contrast matrix $\boldsymbol{\Psi}$. The probabilities in the parameter $\mathbf{p}$ can then be expressed as a function of its coordinates $\mathbf{p}^*$ using the ilr^{-1} transformation

$$p_i = C(\mathbf{p}^*) \exp\left[\sum_{j=1}^{D-1} \psi_{ji} p_j^*\right],$$

where $C(\mathbf{p}^*)$ is the closure constant for $\mathbf{p}$, where its dependence on the coordinates has been enhanced. Substituting this expression in the multinomial probability function yields

$$f_{\mathbf{X}}(\mathbf{x}|\mathbf{p}^*) = \frac{N!}{\prod_{i=1}^{D}(x_i!)} \cdot \exp\left[N \ln\,(C(\mathbf{p}^*)) + \sum_{j=1}^{D-1} p_j^*\left(\sum_{i=1}^{D} x_i \psi_{ji}\right)\right],$$

where the parameters have changed from $\mathbf{p}$ to $\mathbf{p}^*$. This expression of the probability density shows the well-known fact that the multinomial distributions are an exponential family (Witting, 1985; Boogaart et al., 2010). The number of parameters is $D - 1$ as expected, that is, it is a strictly $(D - 1)$-parametric family. The novelty is that here the ilr coordinates appear as the natural parameters of the multinomial family. These facts have some practical consequences, additional to their theoretical value. For instance, the parameters $\mathbf{p}^*$ are no longer constrained to be positive, to add to a constant and they are scaled properly, thus simplifying searching algorithms or integration on the parameters.

Example 7.10 (Estimating probabilities from counts).
An editor in chief of a scientific journal is interested in understanding its impact factor. She obtains a citation report from an indexing service, with regard to the number of citations obtained for each of the papers of an issue (taken at random) of 2 years ago. She considers three classes: papers not cited, papers cited between one and five times, and papers with six or more citations. The report gives a result for that particular issue of 7:4:0. Considering that this corresponds to a multinomial sampling of $N = 11$, the number of papers in that issue, and applying the ML estimator (Equation 7.5) of the probability vector, she would obtain an estimate of $\hat{\mathbf{p}} = [7/11, 4/11, 0]$, that is, the ML estimate assigns zero probability of having a paper with more than five citations. As this is not a realistic result (a different issue could have yield one or two of those more cited papers), she opts for a Bayesian approach. Having no prior conception about the distribution of citations in her journal, she chooses as prior for $\mathbf{p}$ a Dirichlet with $\boldsymbol{\alpha}^0 = [1, 1, 1]$, which is flat with respect to the Lebesgue measure. The posterior is thus also a Dirichlet distribution with $\boldsymbol{\alpha} = [8, 5, 1]$. Figure 7.1 shows this posterior distribution with respect to the Lebesgue and the Aitchison measures, together with two possible point estimates (mean and mode) for each measure. These four statistics are

$$\mathrm{E}_L[\mathbf{X}] = \mathrm{Mode}_a[\mathbf{X}] = \left[\frac{8}{14}, \frac{5}{14}, \frac{1}{14}\right],$$

$$\mathrm{Mode}_L[\mathbf{X}] = \left[\frac{7}{11}, \frac{4}{11}, 0\right],$$

$$\mathrm{E}_a[\mathbf{X}] = C[\exp\,[\psi(8)], \exp\,[\psi(5)], \exp\,[\psi(1)]] = [0.5968, 0.3586, 0.0446].$$

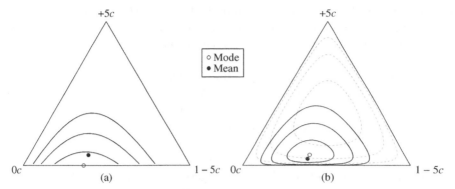

Figure 7.1   Prior (gray dashed lines) and posterior (black solid lines) 50%, 90%, and 99% probability contours of the estimation of proportions of Example 7.10. (a) Results with respect to the Lebesgue measure (prior not shown, as it is uniform). (b) Results with respect to the Aitchison measure.

Figure 7.1 shows as well the uncertainty associated with this estimation, suggesting, for instance, that with 90% probability, one could have found estimates of the proportion of the most cited articles up to 1/3 instead of the naive 0 probability.

After obtaining the Dirichlet posterior with parameter α, it is easy to assess the weight of the prior assumption $\alpha^0 = [1, 1, 1]$. As each 1 in the prior parameter is added to the number of counts in each category, the prior is equivalent to three multinomial trials in which each category is obtained once. Compared with the sample, made of $N = 11$ observed multinomial trials, it means that the editor in chief is confident on her own appreciation (prior) in a ratio $3/11$ relative to the information coming from the sample.   ◇

7.6   Exercises

Exercise 52. Build the subcomposition (Fe_2O_3, FeO, MnO) from the Kilauea data set of Table 5.1, and plot it in a ternary diagram and an ilr scatterplot. Check the normality of this data.

Advanced exercise:

Build the subcompositions $[Fe_2O_3, FeO, MgO]$ and $[Fe_2O_3, FeO, MnO, MgO]$, and test their compositional normality. How can you interpret the result? Which components, or ratios of components, seem to be responsible for a departure from normality?

(Suggestion: Do the advanced exercise with a global test only).

Exercise 53. Build the subcomposition (Fe_2O_3, FeO, MgO) from the Kilauea data of Table 5.1. Split it into two groups (at your will, or better, at random) and perform the different tests of Chapter 7 on the centers and covariance structures.

Exercise 54. Consider the subcomposition (MnO, MgO, CaO) from the Kilauea data of Table 5.1. Split it into two groups depending on whether CO_2 was zero or not, and perform the different tests of normality. If no strong evidence against normality is found, test the equality of centers and covariance structures.

Exercise 55. Compute and plot a 95% confidence region for the ilr-transformed mean of the data from Table 2.1 in $\mathbb{R}^2$. Suggestion: use Section A.2.

Exercise 56. Transform the confidence region of Exercise 55 back into the ternary diagram using the ilr^{-1}.

Exercise 57. Compute and plot a 90% probability region for the ilr-transformed data of Table 2.1 in $\mathbb{R}^2$, together with the sample. Express the ellipse back in compositions and represent it with the data in a ternary diagram. Use the χ^2 distribution, as if mean and variance were actually known.

Exercise 58. Simulate two subpopulations, 50 observations each, of a three-part random composition such that in a basis built with the balances

$$x_1^* = \sqrt{\frac{2}{3}} \ln \frac{x_2^2}{x_1 x_3}, \quad x_2^* = \sqrt{\frac{1}{2}} \ln \frac{x_3}{x_1},$$

these two subpopulation have the same variability and different means:

$$\mu_A = \begin{pmatrix} -1 \\ 0 \end{pmatrix}, \quad \mu_B = \begin{pmatrix} +1 \\ 0 \end{pmatrix}, \quad \Sigma = \begin{pmatrix} 1 & -1/2 \\ -1/2 & 4 \end{pmatrix}.$$

Using the simulated data, test that compositional normality fits each subpopulation. Apply the chain of tests of equal mean and variance, and verify that the hypothesis of equal variances and different means would not be rejected. Plot the two subsets in a ternary diagram (with different colors) and add their respective confidence regions, using the pooled variance.

8

Linear models

Linear models are intended to relate two sets of random variables using linear relationships. They are very general and appear routinely in many statistical applications. A first set of variables, called *response variables*, are to be predicted from a second set of variables, called *predictors* or *covariates*. In the standard approach, linear combinations of the predictors are used to get a predictor function approaching responses, assuming implicitly that the predictors are real random variables obeying Euclidean geometry in real space. In this approach, the predictor function is transformed – if necessary – by some nonlinear function, known as *link function*, and errors or residuals are measured as Euclidean differences between responses and the (transformed) predictor function. There is an extensive literature on general linear models (e.g., Anderson, 1984). The two sets of variables may have very different characteristics (categorical, real, discrete) and the link function choices are also multiple. Here we are interested in cases where responses or predictors have compositional character and we pay special attention to the space where operations are performed. When the response is compositional, we must be aware that residuals should be computed with the perturbation difference defined in page 25, and their size with the Aitchison norm or squared compositional distances, that is, within the framework of the Aitchison geometry of the simplex and not within the framework of the usual Euclidean geometry in real space. Section 8.1 treats the case where the response is compositional and the covariates are real or discrete, thus corresponding to a multiple regression. Response is handled using its coordinates. The ilr^{-1}-transformation plays the role of a link function. Section 8.2 assumes a single real response and a compositional predictor. Again ilr plays an important

Modeling and Analysis of Compositional Data, First Edition.
Vera Pawlowsky-Glahn, Juan José Egozcue and Raimon Tolosana-Delgado.
© 2015 John Wiley & Sons, Ltd. Published 2015 by John Wiley & Sons, Ltd.

role. Section 8.3 discusses the case in which the response is compositional and the predictors reduce to a categorical variable indicating a treatment or a splitting in subpopulations. The goal of such an analysis of variance (ANOVA) model is to decide whether the center of compositions across the treatments are equal or not. Section 8.4 deals with discriminant analysis. In this case, response is a category to which a compositional observation (predictor) is assigned. The model is then a rule to assign an observed composition to categories or treatments. It provides a probability of belonging to each category (see, e.g., Fahrmeir and Hamerle, 1984). This chapter presents the concepts, elements, and results of linear models that get a specific interpretation or expression when used with compositional data. Thus, it does not replace a textbook on multivariate statistics, as those aspects not mentioned here should be considered equal to their conventional analogs. For instance, Bonferroni-type corrections for joint testing, simultaneous tests using Hotellings (T^2)-statistics, or the tables of misclassification of a linear discriminant analysis (LDA) are as valid as within a noncompositional context. Note that models for the case were both responses and predictors have compositional character are not specifically addressed. The straightforward approach consists in taking coordinates separately in each of both sets of samples and proceeds as usual in multiple, multivariate regression.

8.1   Linear regression with compositional response

Linear regression is intended to identify and estimate a linear model from response data that depend linearly on one or more covariates. The assumption is that responses are affected by errors or random deviations of the mean model. The most usual methods to fit the regression coefficients are the well-known least squares techniques.

The problem of regression when the response is compositional is stated as follows. A compositional sample in S^D, denoted by $\mathbf{x}_1, \mathbf{x}_2, \ldots, \mathbf{x}_n$, is available. The sample size is n. Each data point, $\mathbf{x}_i, i = 1, 2, \ldots, n$, is associated with one or more external variables or covariates grouped in the vector $\mathbf{t}_i = [t_{i0}, t_{i1}, \ldots, t_{ir}]$, where $t_{i0} = 1$ by convention. The goal is to estimate the coefficients $\boldsymbol{\beta}_j$ of a curve or surface in S^D with equation

$$\hat{\mathbf{x}}(\mathbf{t}) = (t_0 \odot \boldsymbol{\beta}_0) \oplus (t_1 \odot \boldsymbol{\beta}_1) \oplus \ldots \oplus (t_r \odot \boldsymbol{\beta}_r) = \bigoplus_{j=0}^{r} (t_j \odot \boldsymbol{\beta}_j), \tag{8.1}$$

where $\mathbf{t} = [t_0, t_1, \ldots, t_r]$ are real covariates and are identified as the parameters of the curve or surface to be fitted; again, the first parameter is defined as the constant

$t_0 = 1$, as assumed for the observations before. The compositional coefficients of the model, $\beta_j \in S^D$, will be estimated from the data. The model in Equation (8.1) is very general and includes several useful cases. For instance, polynomial regression on a covariate t is included as a particular case taking $t_j = t^j$. The case of simple regression corresponds to $r = 1$: this is a straight line in the simplex (Equation 9.2).

The most popular fitting method of the model in Equation (8.1) is the least squares deviation criterion. As the response $\mathbf{x}(t)$ is compositional, it is natural to measure deviations also in the simplex using the concepts of the Aitchison geometry. The deviation of the model in Equation (8.1) from the data is defined as $\hat{\mathbf{x}}(t_i) \ominus \mathbf{x}_i$ and its size is measured by the Aitchison norm $\|\hat{\mathbf{x}}(t_i) \ominus \mathbf{x}_i\|_a^2 = d_a^2(\hat{\mathbf{x}}(t_i), \mathbf{x}_i)$. The target function (sum of squared errors, SSE) is

$$\text{SSE} = \sum_{i=1}^{n} \|\hat{\mathbf{x}}(t_i) \ominus \mathbf{x}_i\|_a^2,$$

to be minimized as a function of the compositional coefficients β_j, which are implicit in $\hat{\mathbf{x}}(t_i)$. The number of coefficients to be estimated in this linear model is $(r + 1) \cdot (D - 1)$.

This least squares problem is reduced to $D - 1$ ordinary least squares problems when the compositions are expressed in coordinates with respect to a basis of the simplex. Assume that an orthonormal basis has been chosen in S^D and that the coordinates of $\mathbf{x}_i$, $\hat{\mathbf{x}}(t)$, and β_j are $\mathbf{x}_i^* = [x_{i1}^*, x_{i2}^*, \dots, x_{i,D-1}^*]$, $\hat{\mathbf{x}}^*(t) = [\hat{x}_1^*(t), \hat{x}_2^*(t), \dots, \hat{x}_{D-1}^*(t)]$, and $\beta_j^* = [\beta_{j1}^*, \beta_{j2}^*, \dots, \beta_{j,D-1}^*]$, for $i = 1, 2, \dots, n$ and $j = 1, 2, \dots, r$. All these vectors are in $\mathbb{R}^{D-1}$. As perturbation and powering in the simplex are translated into the ordinary sum and product by scalars in the coordinate real space, the model in Equation (8.1) is expressed in coordinates as

$$\hat{\mathbf{x}}^*(t) = \beta_0^* \, t_0 + \beta_1^* \, t_1 + \dots + \beta_r^* \, t_r = \sum_{j=0}^{r} \beta_j^* \, t_j.$$

For each coordinate, this expression becomes

$$\hat{x}_k^*(t) = \beta_{0k}^* \, t_0 + \beta_{1k}^* \, t_1 + \dots + \beta_{rk}^* \, t_r, \quad k = 1, 2, \dots, D - 1. \tag{8.2}$$

Also, the Aitchison norm and distance become the ordinary norm and distance in real space. Then, using orthonormal coordinates, the target function is expressed as

$$\text{SSE} = \sum_{i=1}^{n} \|\hat{\mathbf{x}}^*(t_i) - \mathbf{x}_i^*\|^2 = \sum_{k=1}^{D-1} \left\{ \sum_{i=1}^{n} |\hat{x}_k^*(t_i) - x_{ik}^*|^2 \right\}, \tag{8.3}$$

where $\| \cdot \|$ is the norm of a real vector. All sums in Equation (8.3) are nonnegative and, therefore, the minimization of SSE implies the minimization of each term of the sum in k,

$$\text{SSE}_k = \sum_{i=1}^{n} |\hat{x}_k^*(t_i) - x_{ik}^*|^2, \quad k = 1, 2, \ldots, D - 1.$$

That is, fitting of the compositional model in Equation (8.1) reduces to the $D - 1$ ordinary least squares problems in Equation (8.2). From this point on, we can use standard regression tools and techniques (coefficient confidence intervals, individual tests, residuals, predictions, etc.) on each coordinate individually. Pooling results from all coordinates together, those of these concepts that identify again a composition (coefficients, residuals, and predictions) can be represented back by the inverse of the ilr used.

Example 8.1 (Religions in Brazil).
Table 8.1 shows the percentages of Brazilians adhering to Catholicism, Evangelist churches, other religions, and no religion at all. The same table also shows the simplest results of a regression analysis. The first part of the table (upper left)

Table 8.1   Evolution of religious beliefs in Brazil since 1940. Data in percentage. Data from CPS/FGV, provided by IBGE (Fundaçao Gelulio Vargas). Legend: C, Catholic; E, Evangelist; O, other; N, no religion; R, all religions.

Year	Time	C	E	O	N	C/E	CE/O	R/N
1940	0	96.06	2.63	0.81	0.51	2.54	2.43	2.13
1950	10	94.92	3.45	0.81	0.81	2.34	2.53	1.79
1960	20	94.32	4.05	1.01	0.61	2.23	2.42	2.15
1970	30	92.91	5.26	1.01	0.81	2.03	2.52	1.97
1980	40	91.09	6.76	0.92	1.23	1.84	2.69	1.65
1991	51	84.66	9.14	1.02	5.18	1.57	2.70	0.50
2000	60	73.85	16.17	2.60	7.39	1.07	2.11	0.59
Intercept[†]	35.44	0.87	0.25	0.13	2.62	2.52	2.37	
Slope[†]	−0.0243	0.0075	−0.0070	0.0238	−0.0224	−0.0011	−0.0275	
Intercept[‡]	96.60	2.36	0.69	0.35	0.0002[*]	0.8[*]	0.02[*]	
Slope[‡]	0.9760	1.0075	0.9930	1.0241				

[*] Significance of the slopes (see text for details), only the first nonzero figure is shown.
[†] Logratio intercept and slope: the right table reports these coefficients in ilr, while the left table reports them in clr.
[‡] Compositional intercept and slope: the intercept is shown as a closed composition adding up to 100%, while the slope is a vector of multiplicative coefficients.

shows the data, as percentages of each belief every 10 years since 1940. Time is recomputed as the years passed since the first register. The second part of the table (upper right) provides the ilr-coordinates of this data set, in a hierarchical partition: first Catholics against Evangelists, then these two Christian confessions against other religions, and finally religious against nonreligious. The third part of the table reports the classical least squares intercept and slope estimates for each one of the coordinates as a function of time. The intercept thus represents the (conditional) average logratios in 1940. This intercept, taken as vector of coordinates, can be ilr-backtransformed to provide the (conditional) average percentages of beliefs in 1940. This is the content of the fourth part of the table (lower right): intercept and slope recast in terms of the original four parts.

Results show that all chosen balances decrease with time, that is, along time, the relative importance of Catholics against Evangelists decreases, as happens with the relative importance of these two Christian confessions combined against other religions. But also the relative importance of (a combination of) all religious beliefs against nonreligious beliefs decreases. How these different trends combine together? It is obvious that nonreligious beliefs increase their relative importance over the period, as they are involved in only one balance in the denominator. It is also easy to see that Catholics loose relative importance every year, as they are always in the numerator of all balances. But what happens with the other two variables? We can see it recasting the vector of three slopes from its current representation as an ilr vector to a clr vector, using Equation (4.8). With this operation, we may conclude that Evangelists (in relation to all other three beliefs) tend to increase, while other religions tend to decrease. The pattern for Catholics and nonreligious was already clear, but as clr's we see again that the relative proportion of the first will tend to decrease while that of the last will increase.

The slope of this regression trend is interpreted as a relative rate of increase/decrease of the *proportion* of each belief. Evolution of the *actual* number of people belonging to each belief may show a completely different pattern. It depends on how the number of inhabitants in Brazil evolves. It actually happened that the Brazil population exploded in the past 60 years, and the numbers of followers of all religions increased, although their proportions show the trends explained before, because the number of nonreligious people has increased much faster. To model this effect, we need data (or a model) for the evolution of the total population, which is beyond the scope of this text. A hint on how it could be worked out can be found in Chapter 10. The three intercepts of these regression models can be combined together and the inverse ilr may be applied to them. The result is also reported in Table 8.1. This shows an estimate of the proportions of each belief in 1940 (the origin of the series): obviously, Catholics were the overwhelming majority, while the proportions of other beliefs were extremely low.

To assess the significance of trends, we can derive a global goodness-of-fit measure, which is in the present case $R^2 = 0.988$. Some statistical software (for instance, R) may be able to derive a multivariate goodness-of-fit test on this joint R^2. The associated p-value will give the significance of the null hypothesis that $R^2 = 0$. In this case, R yields a p-value of 0.0009367, much lower than the typical 5% critical value used for not rejecting the null hypothesis. Thus, we can be confident that the trend described, as a whole, is significant.

There is still the question of which of the ratios do actually evolve in time. For instance, it could happen that the relative proportion of Evangelist to other religions is more or less constant in time. To solve this, we can easily compute the associated balance coordinates in CoDaPack, Excel, or any spreadsheet software and then use a statistical package to fit a regression to each coordinate and check the significance of the coefficients. Table 8.1 also reports the significance of this coordinate slope. It shows that the most significant trend is the relative decrease of Catholics against Evangelists (p-value $= 2.324 \cdot 10^{-4} < 0.01$), followed by the general decrease of religious people relative to nonreligious people (p-value $= 1.555 \cdot 10^{-2} < 0.05$). The second balance shows a nonsignificant trend (p-value $= 0.791 > 0.1$), which implies a certain equilibrium in the relative increase of other religions against the evolution of Evangelists and Catholics (Figure 8.1).

Finally, the model can be used to derive proportions of these beliefs for times not in the table, either within the data set time span (e.g., for 1985) or in the (near) future or past (1900 or 2025). For each desired time t, one simply applies the formula $y = a + b \cdot t$ for a the intercept and b the slope and obtains a prediction of a given coordinate. Predictions of the three coordinates can be backtransformed with the inverse ilr, to obtain a prediction of the proportions themselves. This procedure has been applied to obtain the lines of Figure 8.1, which also displays the data for comparison.   ◇

Example 8.2 (Vulnerability of a system).
A system faces external actions. Its response to such actions is frequently a major concern in risk management and environmental impact assessment. For instance, the system may be a dike under the action of ocean-wave storms. Or the system may be a nuclear plant, and the action an earthquake or a tsunami. Or the system may be a wetlands area, and the action a weather-induced drought or an industrial pollution event. In all cases, the response would be the level of service or health of the system after one event. We consider a dike versus wave scenario as example. Three responses of the system may be considered: θ_1, service (healthy); θ_2, damage (but recoverable); θ_3 collapse (unrecoverable). The dike can be designed for a design action, for example, wave-height, d, with values $3 \leq d \leq 20$ (wave-height

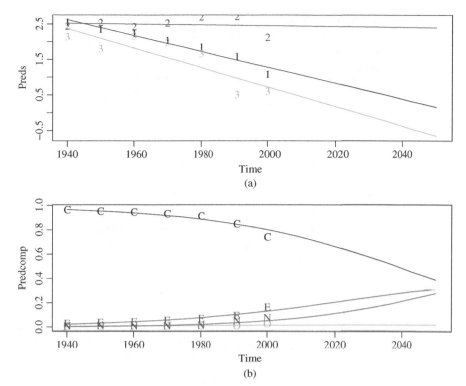

Figure 8.1   Evolution diagrams of (a) the balances and (b) the proportions. Numbers and letters represent the data: C, Catholics; E, Evangelists; O, other religions; N, nonreligious; 1, C/E balance; 2, CE/O balance; 3, CEO/N balance.

in meters). Actions are parameterized by some wave-height of the storm, h, also with values $3 \leq d \leq 20$ (wave-height in meters). Vulnerability of the system is described by the conditional probabilities

$$p_k(d, h) = \Pr[\theta_k | d, h], \quad k = 1, 2, 3 = D, \quad \sum_{k=1}^{D} p_k(d, h) = 1,$$

where, for any pair (d, h), $\mathbf{p}(d, h) = [p_1(d, h), p_2(d, h), p_3(d, h)] \in S^3$. In practice, $\mathbf{p}(d, h)$ is only approximately known for a limited number of values $\mathbf{p}(d_i, h_i)$, $i = 1, \dots, n$. The whole model of vulnerability can be expressed as a regression model

$$\hat{\mathbf{p}}(d, h) = \boldsymbol{\beta}_0 \oplus (d \odot \boldsymbol{\beta}_1) \oplus (h \odot \boldsymbol{\beta}_2), \tag{8.4}$$

so that it can be estimated by regression in the simplex.

Table 8.2   Assumed vulnerability for a dike with only three outputs or responses. Probability values of the response θ_k conditional to values of design d and level of the storm h.

d_i	h_i	Service	Damage	Collapse
3.0	3.0	0.50	0.49	0.01
3.0	10.0	0.02	0.10	0.88
5.0	4.0	0.95	0.049	0.001
6.0	9.0	0.08	0.85	0.07
7.0	5.0	0.97	0.027	0.003
8.0	3.0	0.997	0.0028	0.0002
9.0	9.0	0.35	0.55	0.01
10.0	3.0	0.999	0.0009	0.0001
10.0	10.0	0.30	0.65	0.05

Consider the data in Table 8.2 containing $n = 9$ probabilities. Figure 8.2 shows the vulnerability probabilities obtained by regression for six design values. An inspection of these figures reveals that a quite realistic model has been obtained from a really poor sample: service probabilities decrease as the level of action increases and conversely for collapse. This changes smoothly for increasing design level, so that, typically, the system resists actions under the design level. Despite the challenging shapes of these curves describing the vulnerability, they come from a linear model, as can be seen in Figure 8.3a. In Figure 8.3b these straight lines in the simplex are shown in a ternary diagram.   ◇

8.2   Regression with compositional covariates

The model with compositional covariates appears when the goal is to predict one external variable as a function of a composition. Assume a compositional data set, $x_1, x_2, \ldots, x_n$, is available and that each data point x_i is associated with an observation y_i of an external response variable (with support on the *whole real line*, i.e., possibly transformed with logs or any other suitable transformation). The goal is to estimate a surface on $S^D \times \mathbb{R}$ with equation

$$\hat{y}(x) = \beta_0 + \langle \beta, x \rangle_a, \tag{8.5}$$

where $\beta \in S^D$ is the (simplicial) gradient of y with respect to x and β_0 is a real intercept. Note that the gradient β is introduced in the Aitchison inner product

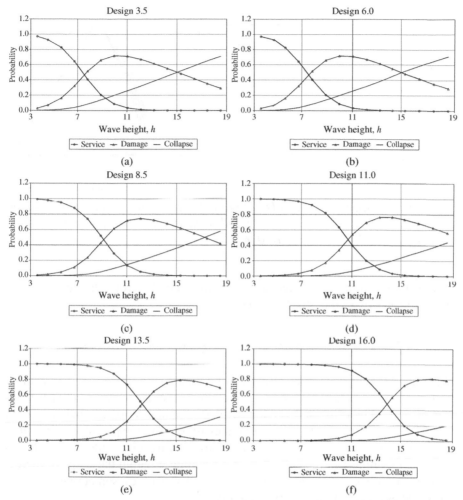

Figure 8.2   Vulnerability models obtained by regression in the simplex from the data in Table 8.2. Horizontal axis: incident wave-height in meters. Vertical axis: probability of the output response. Shown designs are 3.5, 6.0, 8.5, 11.0, 13.5, 16.0 (design wave-height in meters).

with **x**, whose result is a real number. In the present case, because the response is a real value, the classical least squares fitting criterion may be applied, which yields the target function

$$\text{SSE} = \sum_{i=1}^{n} (y_i - \beta_0 - \langle \boldsymbol{\beta}, \mathbf{x}_i \rangle_a)^2 \ .$$

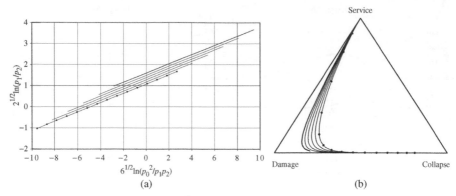

Figure 8.3   Vulnerability models in Figure 8.2 (a) in coordinates and (b) in the ternary diagram. Circles: Design 3.5. Subsequent lines correspond to design levels 6.0, 8.5, 11.0, 13.5, and 16.0 m.

As the Aitchison inner product can be computed easily from clr coefficients or ilr coordinates of $\mathbf{x}_i$, the SSE becomes

$$\text{SSE} = \sum_{i=1}^{n} (y_i - \beta_0 - \langle \text{clr}(\boldsymbol{\beta}), \text{clr}(\mathbf{x}_i)\rangle)^2 = \sum_{i=1}^{n} (y_i - \beta_0 - \langle \text{ilr}(\boldsymbol{\beta}), \text{ilr}(\mathbf{x}_i)\rangle)^2. \quad (8.6)$$

This suggests that the actual fitting can be done using ilr coordinates without further ado. One simply fits a linear regression to the response y as a linear function of ilr($\mathbf{x}$). The estimated version of ilr($\boldsymbol{\beta}$) contains slope coefficients of the response y with respect to the coordinates ilr($\mathbf{x}$). The simplicial gradient $\boldsymbol{\beta}$ is then easily computed using ilr^{-1}. The clr-transformation should be avoided in this case for numerical reasons: correctly handling it within the scope of regression requires the generalized inversion of singular matrices, something that most statistical packages are not prepared to do.

Tests on the coefficients of the model in Equation (8.5) may be used as usual, for instance, to obtain a simplified model depending on less coordinates. However, one should be aware that they are related to the particular basis used and that these simplified models will be different depending on the basis. If simplification is needed, one should carefully select the basis: for instance, a basis of balances may be adequate to check the dependence of y on a particular subcomposition of $\mathbf{x}$. In particular, if there are reasons to suspect that some components of $\mathbf{x}$ might not be influential on y. To detect which components might be superfluous and choose a good basis to test, one can fit a regression model using any ilr transformation. This provides a gradient in ilr coordinates ilr($\hat{\boldsymbol{\beta}}$), which can be reexpressed as a gradient in clr coefficients clr($\hat{\boldsymbol{\beta}}$) = $\hat{\mathbf{b}}$ by Equation (4.8). As a general principle, the ideal ilr basis for this problem should group together those components with a similar $\hat{b}_i$, as shown in Example 8.3.

Example 8.3 (Religions in Brazil revisited).
This example revisits the same data set of religious beliefs proportions in Brazil of Example 8.1. Although it is not the common way of understanding it, we may try to predict the year at which certain proportions of religions are likely to occur. For that, we want to obtain a model of *y* (*years after 1940*) as a function of the beliefs proportions,

$$\hat{y}(\mathbf{x}) = \beta_0 + \beta_1^* x_1^* + \beta_2^* x_2^* + \beta_3^* x_3^*,$$

where $\mathbf{x}^* = [x_1^*, x_2^*, x_3^*] = \mathrm{ilr}(\mathbf{x})$ is the ilr-transformed beliefs proportions and $\boldsymbol{\beta}^* = [\beta_1^*, \beta_2^*, \beta_3^*] = \mathrm{ilr}(\boldsymbol{\beta})$ the ilr-transformed gradient. This can be fitted with any software apt for multivariate regression. Using the same ilr basis as before, and standard R regression routines, we obtain the coefficients (with their significance) shown in Table 8.4. It can be seen that, apparently, the last coordinate, ln (CatholicEvangelistOther/No.religion), can be removed from the model. This implies that only the subcomposition (Catholic, Evangelist, Other) is relevant for time prediction.

Alternatively, we might recast the resulting ilr gradient to clr, which gives $[C, E, O, N] = [-24.9, 50.2, -20.0, -5.3]$. Though the numbers are quite different, the coefficients of Catholics and other religions are the most similar, followed by the nonreligious. This suggests that using a partition such as the one given in Table 8.3, where v_2 will compare Catholics versus other religions: given that the coefficients of the clr-gradient for these two were similar, it is possible that the regression model expressed in the ilr basis associated with this partition has a zero coefficient in this particular coordinate. This is confirmed in Table 8.4 with a conventional test on the coefficients. Thus, to predict the year as a function of the proportions between religions, one has enough using the relation E/N and the balance (CO)/(NE).   ◇

Table 8.3   Sequential binary partition of proportions of religions in Brazil

	C	E	O	N
v_1	+1	−1	+1	−1
v_2	+1	0	−1	0
v_3	0	−1	0	+1

Legend: C, Catholic; E, Evangelist; O, other; N, no religion.

Table 8.4   Regression estimates, coefficient estimation errors, and p-values of the null-coefficient hypotheses for the prediction of years after 1940 as a function of religions proportions in Brazil, using two different bases.

	Estimate	Standard error	t value	$\Pr(> |t|)$
(Intercept)	57.716	13.042	4.425	0.021
C vs. E	−53.132	4.834	−10.992	0.002
CE vs. O	26.715	5.846	4.570	0.020
CEO vs. N	6.138	3.292	1.864	0.159
CO vs. NE	−44.933	3.676	−12.222	0.001
C vs. O	−3.430	4.523	−0.758	0.503
N vs. E	−39.289	5.865	−6.699	0.007

8.3   Analysis of variance with compositional response

ANOVA is the name given to a linear model where a continuous response is explained as a function of a (set of) discrete variable(s). Compositional ANOVA follows the same steps that were used to predict a composition from a continuous covariate. Notation in multiway ANOVA (with more than one discrete covariate) can become quite cumbersome, thus only one-way compositional ANOVA will be addressed here. Textbooks on multivariate ANOVA may be then useful to extend this material. Readers interested in compositional multiway ANOVA are referred to Boogaart and Tolosana-Delgado (2013).

As in the preceding section, it is assumed that a compositional sample $x_1, x_2, \ldots,$ x_n in S^D is available. These observations are classified into K categories across an external categorical variable z. The variable z may represent different treatments or subpopulations. In other words, for each composition, a category z_i is given. ANOVA deals with the centers (compositional means) of the composition for each category, $\mu_1, \ldots, \mu_K$. Following classical notation, a compositional ANOVA model, for a given z, is

$$\hat{x} = \beta_1 \oplus (I(z = 2)) \odot \beta_2) \oplus \ldots \oplus (I(z = K) \odot \beta_K), \quad x \ominus \hat{x} = \epsilon,$$

where the indicator $I(z = k)$ equals 1 when the condition is true and 0 otherwise. This means that only one of the $K - 1$ terms may be taken into account for each possible z value (as the other terms will all be powered by zero). Note that the first category does not explicitly appear in the equation: when $z = 1$, then the predictor is only $\beta_1 = \mu_1$, the center of the first group. The remaining coefficients are then interpreted as increments of the mean composition from the reference first level to each category, $\boldsymbol{\mu}_k = \boldsymbol{\beta}_1 \oplus \boldsymbol{\beta}_k$ or, equivalently, $\boldsymbol{\beta}_k = \boldsymbol{\mu}_k \ominus \boldsymbol{\mu}_1$. Finally, $\boldsymbol{\epsilon}$ is the compositional residual of the model.

ANOVA notation may be quite cumbersome, but its fitting is straightforward. For each value z_i, a vector $\mathbf{t}_i$ with the $K - 1$ indicators is constructed, that is, a vector of zeroes except for a single 1 in the position of the observed category. With these vectors, the same steps as with linear regression with compositional response (Section 8.1) are applied. An important consequence of this procedure is that implicitly an equal total variance within each category is assumed (compare with Chapter 7).

The practice of compositional ANOVA follows the principle of working on coordinates. First, a basis is selected and the corresponding coordinates of the observations are computed. Then classical ANOVA is applied to explain each ilr coordinate as a mean reference level plus $K - 1$ mean differences between categories. Tests may be applied to conclude that some of these differences are not significant, that is, the mean of a particular coordinate at two categories may be taken as equal. Finally, all coefficients associated with category k can be jointly backtransformed using ilr^{-1} to obtain its associated compositional coefficient $\boldsymbol{\beta}_k$.

Example 8.4 (Nutrition (continued)).
Following the example in Section 5.8, the question is now to evaluate the global differences between eastern and western countries. To proceed with an ANOVA of the composition against this categorical variable, first an ilr basis needs to be selected. That choice should be guided by previous knowledge. But if it is not available, it is possible to proceed as follows. First a black-box ilr is used, based on the following hierarchical binary partition:

	RM	WM	E	M	F	C	S	N	V
v_1	−1	1	0	0	0	0	0	0	0
v_2	−1	−1	1	0	0	0	0	0	0
v_3	−1	−1	−1	1	0	0	0	0	0
v_4	−1	−1	−1	−1	1	0	0	0	0
v_5	−1	−1	−1	−1	−1	1	0	0	0
v_6	−1	−1	−1	−1	−1	−1	1	0	0
v_7	−1	−1	−1	−1	−1	−1	−1	1	0
v_8	−1	−1	−1	−1	−1	−1	−1	−1	1

Readers may find it interesting to construct the associated $\boldsymbol{\Psi}$ matrix before continuing. Once the matrix is available, a data set of ilr coordinates is obtained. Then, any statistical software can fit an ANOVA model to each of these coordinates separately. This yields for the data considered the following vectors of coefficients:

	v_1	v_2	v_3	v_4	v_5	v_6	v_7	v_8
β_1^*	−0.08	−1.08	0.82	−1.40	2.11	−0.60	−0.57	−0.36
β_2^*	−0.20	0.27	0.11	0.92	−0.96	−0.10	−0.65	−0.17

which can be multiplied by Ψ to obtain the estimated clr coefficients:

	RM	WM	E	M	F	C	S	N	V
$\mathrm{clr}\hat{\beta}_1$	0.40	0.28	−0.98	0.85	−1.43	2.14	−0.44	−0.49	−0.34
$\mathrm{clr}\hat{\beta}_2$	0.09	−0.19	0.28	0.19	1.12	−0.76	0.01	−0.59	−0.16

Now, $\mathrm{clr}\hat{\beta}_2$ is describing the difference from East to West, that is, a positive coefficient means that this particular component was more consumed in Western countries than in Eastern countries, with regard to all other components (as it is a clr). On themselves, these coefficients can be misleading, for two reasons: (i) each clr coefficient depends on the whole set of components and (ii) it is not known how statistically significant these coefficients are. The idea is to use them as a guide to select an ilr basis, to which statistical tests can be applied. This can be done by grouping together those components with a similar clr-coefficient, that is, white meat with vegetables, red meat with starch, eggs with milk, and nuts with cereals. These can then be grouped further in a hierarchy, for example,

	RM	WM	E	M	F	C	S	N	V
v_1'	0	−1	0	0	0	0	0	0	1
v_2'	−1	0	0	0	0	0	1	0	0
v_3'	0	0	−1	1	0	0	0	0	0
v_4'	0	0	0	0	0	−1	0	1	0
v_5'	−1	0	1	1	0	0	−1	0	0
v_6'	1	−1	1	1	0	0	1	0	−1
v_7'	1	1	1	1	0	−1	1	−1	1
v_8'	1	1	1	1	−1	1	1	1	1

Again, the reader is recommended to build the associated Ψ' matrix. An ANOVA of the coordinates with respect to this tailored basis will probably

Table 8.5   ANOVA coefficients, marginal errors, and significance of the zero-coefficient hypothesis for the European nutrition data patterns, in the tailored basis

	Estimate	Standard error	t value	$\Pr(> |t|)$
v_1'	−0.20	0.18	−1.09	0.285
v_2'	0.27	0.10	2.82	0.010
v_3'	0.11	0.13	0.87	0.391
v_4'	0.92	0.32	2.92	0.008
v_5'	−0.96	0.19	−5.12	0.000
v_6'	−0.10	0.17	−0.60	0.556
v_7'	−0.65	0.35	−1.87	0.075
v_8'	−0.17	0.18	−0.98	0.336

yield several coefficients that can be removed from the model because of a high p-value of the hypothesis of being zero (Table 8.5). With this table in mind, coordinates v_6', v_3', v_8', v_1', and v_7' can be discarded to have any significant difference between Eastern and Western countries. If only those variables with a p-value smaller than 5% are retained, Western and Eastern countries differ only in the balances: milk−eggs versus red meat−starch (negative coefficient, i.e., relatively less consumption of the first against the second in Western countries than in the Eastern ones), nuts versus cereals (positive coefficient, i.e., relatively more nut consumption with regard to cereals in Western countries), and starch versus red meat (positive coefficient, i.e., higher meat/starch ratio in the East than in the West). At this 5% critical level, fish, white meat, and vegetables do not show significant differences between Eastern and Western countries, as the balances involving them were all removed because of having high p-values.   ◇

8.4   Linear discrimination with compositional predictor

A composition can also be used to predict a categorical variable. Following notation in the preceding section, the goal is now to estimate $\mathbf{p}(\mathbf{x})$, the probability that z takes each of K possible values given an observed composition $\mathbf{x}$. There are many techniques to obtain this result, such as the Fisher rule, linear or quadratic discriminant analysis, or multinomial logistic regression. In any case, this is straightforward by applying the principle of working on coordinates (taking ilr's, applying a method to the coordinates, and backtransforming the coefficients to interpret them as compositions). This Section illustrates this procedure with

LDA, because this technique is available in most basic statistical packages and has a simple geometric interpretation in the simplex.

First, LDA assumes some prior probabilities p_k^0 of data points corresponding to each one of the K categories. These are typically taken as $p_k = 1/K$, $k = 1, 2, \ldots, K$, (equally probable groups) or $p_k = n_k/n$ (where n_k is the number of samples in the kth category). Then, LDA assumes that the ilr-transformed composition has a normal distribution, with ilr mean $\boldsymbol{\mu}_k^* = \mathrm{ilr}(\boldsymbol{\mu}_k)$ and common ilr-coordinate covariance matrix $\boldsymbol{\Sigma}$ (i.e., all categories have the same covariance and possibly different mean, a setting parallel to the ANOVA of Section 8.3). Applying Bayes' formula, the posterior probability vector of belonging to each class for a particular composition $\mathbf{x}$ can be derived from the discriminant functions

$$d_{jk}(\mathbf{x}) = \ln \frac{p_j}{p_k} = A_{jk} + (\boldsymbol{\mu}_j^* - \boldsymbol{\mu}_k^*) \cdot \boldsymbol{\Sigma}^{-1} \cdot \mathbf{x}^{*\mathsf{T}},$$

with

$$A_{jk} = \ln \frac{p_j^0}{p_k^0} - \frac{1}{2}(\boldsymbol{\mu}_j^* - \boldsymbol{\mu}_k^*) \cdot \boldsymbol{\Sigma}^{-1} \cdot (\boldsymbol{\mu}_j^* - \boldsymbol{\mu}_k^*)^{\mathsf{T}}.$$

Again, as happened with ANOVA, one category is typically placed as a sort of reference level. For instance, taking $j = K$ fixed, LDA computes the logodds of the other $(K - 1)$ categories with respect to the last one. The desired probabilities can be obtained with the inverse alr transformation, as explained in Section 4.7.

Obtained probabilities can then be used to decide which category is more likely for each possible composition $\mathbf{x}$: typically, each point in S^D is classified into the most likely group, the one with largest probability. In this sense, the discriminant functions can be used to draw the boundaries between regions j and k, by identifying the set of points where $d_{jk}(\mathbf{x}) = 0$. Some linear algebra shows that this boundary is the affine hyperplane of S^D orthogonal to the vector $\mathbf{v}_{dj}$ and passing through the point $\mathbf{x}_{jk}^0$ obtained as

$$\mathbf{v}_{jk} = (\boldsymbol{\mu}_j^* - \boldsymbol{\mu}_k^*)\boldsymbol{\Sigma}^{-1},$$

$$\mathbf{x}_{jk}^0 = \frac{1}{2} \left[\boldsymbol{\mu}_j^* + \boldsymbol{\mu}_k^* - \frac{A_{ij}}{\langle \boldsymbol{\mu}_j^* - \boldsymbol{\mu}_k^*, \mathbf{v}_{jk} \rangle} \cdot (\boldsymbol{\mu}_j^* - \boldsymbol{\mu}_k^*) \right].$$

Note that these equations are useful only to draw boundaries between neighboring categories, that is, between the two most probable categories of a given point in S^D. For more than two categories, care should be taken to draw them by segments.

Example 8.5 (Nutrition (continued)).
In Example 8.4, it was shown that knowing whether a certain country lies in Western or Eastern Europe, certain differences on consumption of nuts, cereals,

red meat, starch, milk, and eggs are to be expected. The question is now reversed: given a certain nutrition pattern of a country, is it possible to guess at an Eastern or a Western country? For that, the contrast matrix Ψ' of Example 8.4 is selected, and the ilr coordinates of all countries with respect to it are calculated. To avoid over-fitting, only those coordinates that resulted significative in the ANOVA example are used. Importing those coordinates into a suitable software, an LDA is applied, yielding the following results:

- prior probabilities, $p_E = 0.36$ and $p_W = 0.64$, corresponding to, respectively, 9 and 16 countries in each group;

- group means in coordinates, and a discriminating direction.

	v_2'	v_4'	v_5'
μ_E^*	−1.08	−1.40	2.11
μ_W^*	−0.81	−0.48	1.14
$\mathbf{v}_{jk}$	0.88	0.47	−1.75

These result in probabilities of belonging to the Western or the Eastern blocks (Table 8.6). As can be seen, all Western countries were correctly classified, while three Eastern countries (East Germany, Poland, and Czechoslovakia) had higher probabilities of belonging to the Western block than to the Eastern block. Note, nevertheless, the small difference between probabilities that made Czechoslovakia fall in the Western block. This result is not very surprising, as the food habits of these three countries and Western Germany were probably very similar in the subcomposition defined by nuts, cereals, red meat, starch, milk, and eggs, those parts involved in the three balances used.   ◇

8.5   Exercises

Exercise 59. Consider the data in Table 8.7. They are sand–silt–clay compositions from an Arctic lake taken at different depths (adapted from Coakley and Rust (1968) and reproduced in Aitchison (1986)). The goal is to check whether there is some trend in the composition related to depth. Particularly, using the standard hypothesis testing in regression, check the constant and the straight line models

$$\hat{\mathbf{x}}(t) = \boldsymbol{\beta}_0, \quad \hat{\mathbf{x}}(t) = \boldsymbol{\beta}_0 \oplus (t \odot \boldsymbol{\beta}_1),$$

being t = depth. Plot both models, the fitted model and the residuals, in coordinates and in the ternary diagram. Try also with depth logarithmically transformed.

Table 8.6   Probabilities of belonging to the Eastern and Western blocks of each country, according to their consumption patterns in the coordinates v_2', v_4', and v_5', compared with their true group.

Country	p_E	p_W	Group Predicted	True
Albania	1.00	0.00	E	E
Austria	0.11	0.89	W	W
Belgium	0.03	0.97	W	W
Bulgaria	0.98	0.02	E	E
Czechoslovakia	0.47	0.53	W	E
Denmark	0.00	1.00	W	W
Eastern Germany	0.02	0.98	W	E
Finland	0.02	0.98	W	W
France	0.04	0.96	W	W
Greece	0.13	0.87	W	W
Hungary	0.97	0.03	E	E
Ireland	0.05	0.95	W	W
Italy	0.23	0.77	W	W
Netherlands	0.05	0.95	W	W
Norway	0.01	0.99	W	W
Poland	0.29	0.71	W	E
Portugal	0.08	0.92	W	W
Romania	0.97	0.03	E	E
Spain	0.03	0.97	W	W
Sweden	0.00	1.00	W	W
Switzerland	0.13	0.87	W	W
United Kingdom	0.02	0.98	W	W
USSR	0.56	0.44	E	E
Western Germany	0.01	0.99	W	W
Yugoslavia	0.99	0.01	E	E

Exercise 60. Following Exercise 59, one can equivalently check whether the sediment composition brings any information on the depth at which that sample was taken. Using the data from Table 8.7, fit a linear model to explain depth as a function of the composition. Analyze the residuals as usual, as they may be considered real values.

To display the model, you can follow these steps. Split the range of observed depths in several segments of the same length (four to six will be enough in CoDaPack), and give each sample a number corresponding to its depth category.

Table 8.7   Sand, silt, clay composition of sediment samples at different water depths in an Arctic lake

Sample no.	Sand	Silt	Clay	Depth (m)	Sample no.	Sand	Silt	Clay	Depth (m)
1	77.5	19.5	3.0	10.4	21	9.5	53.5	37.0	47.1
2	71.9	24.9	3.2	11.7	22	17.1	48.0	34.9	48.4
3	50.7	36.1	13.2	12.8	23	10.5	55.4	34.1	49.4
4	52.2	40.9	6.9	13.0	24	4.8	54.7	40.5	49.5
5	70.0	26.5	3.5	15.7	25	2.6	45.2	52.2	59.2
6	66.5	32.2	1.3	16.3	26	11.4	52.7	35.9	60.1
7	43.1	55.3	1.6	18.0	27	6.7	46.9	46.4	61.7
8	53.4	36.8	9.8	18.7	28	6.9	49.7	43.4	62.4
9	15.5	54.4	30.1	20.7	29	4.0	44.9	51.1	69.3
10	31.7	41.5	26.8	22.1	30	7.4	51.6	41.0	73.6
11	65.7	27.8	6.5	22.4	31	4.8	49.5	45.7	74.4
12	70.4	29.0	0.6	24.4	32	4.5	48.5	47.0	78.5
13	17.4	53.6	29.0	25.8	33	6.6	52.1	41.3	82.9
14	10.6	69.8	19.6	32.5	34	6.7	47.3	46.0	87.7
15	38.2	43.1	18.7	33.6	35	7.4	45.6	47.0	88.1
16	10.8	52.7	36.5	36.8	36	6.0	48.9	45.1	90.4
17	18.4	50.7	30.9	37.8	37	6.3	53.8	39.9	90.6
18	4.6	47.4	48.0	36.9	38	2.5	48.0	49.5	97.7
19	15.6	50.4	34.0	42.2	39	2.0	47.8	50.2	103.7
20	31.9	45.1	23.0	47.0					

Plot the compositional data in a ternary diagram, using colors for each depth interval. Draw a line on the simplex, from the center of the data set along the gradient of the fitted model.

Exercise 61. Following Example 8.3, we want to choose a basis that allows for the maximal reduction on the number of coordinates necessary for prediction. For this goal,

- write down the matrix Ψ of three rows and four columns that was used to calculate the balances of Table 8.4;

- take the vector of ilr slope coefficients $\beta^* = [-53.132, 26.715, 6.138]$ and express them as clr slope coefficients, that is, calculate $\beta^* \Psi = \beta^*_{clr}$;

- considering the signs and magnitudes of the resulting clr coefficients, choose a hierarchical partition isolating first the quasi-zero components from the other components and then splitting the nonzero components between positive and negative components; within each group (quasi-zero, positive, negative), you can choose how to split the components; write the associated contrast matrix Ψ';

- calculate the coordinates with respect to this basis;

- fit the model again using the new set of coordinates as explanatory variables; test the significance of each logratio;

- recast the clr-coefficients to the new set of balances, with $\beta^*_{clr} \cdot (\Psi')^\mathsf{T}$; compare with the slopes obtained in the preceding point.

Given these results (in particular, the several p-values), can any component be considered to be superfluous to predict the year?

Advanced exercise:

Consider the possibility to restrict the dependence of year to one single coordinate. Which coordinate should that be? Build the corresponding model, and calculate the predicted times $\hat{y}(\mathbf{x})$. Calculate the SSE (Equation 8.6) using these predictions and the predictions provided by the general model (without simplifications, the one with the three slopes of the main exercise). Denote these two SSEs as s_0^2 and s_g^2. Using the generalized likelihood ratio test of Equation (7.6), test the hypothesis that the two models are equivalent, that is, obtain the logratio

$$Q(\mathbf{x}, y) = 2 \ln \frac{s_0^2}{s_g^2}.$$

Obtain the p-value of Q according to a χ^2 distribution of 2 degrees of freedom, the number of restrictions introduced passing from a model with three predictor coordinates to one coordinate. This can be done, for example, in excel with the function CHIINV or in R with function pchisq(..., lower.tail=FALSE). Given this p-value, is there any reason to reject the model with one single predictor coordinate?

Exercise 62. Following Example 8.4, compute the center (Definition 5.1) of the nutrition data set separately for the Eastern and Western countries. Calculate the perturbation difference between them. Compare the two centers and the perturbation with $\hat{\beta}_1$ and $\hat{\beta}_2$ as obtained in Example 8.4 (note that you will have to apply the inverse clr transformation to obtain these estimates first).

Exercise 63. Repeat Example 8.4 and check for differences between Northern and Southern countries.

Table 8.8   Main anion composition of some water samples from four different rivers in Barcelona province (NE Spain).

River	Cl	SO_4	HCO_3	River	Cl	SO_4	HCO_3
A	197.43	857.99	348.39	U	16.54	71.88	182.20
A	312.37	487.83	377.13	U	27.29	93.35	197.97
A	15.49	239.93	146.00	U	26.00	96.81	176.96
A	118.09	445.63	341.50	U	29.15	76.87	188.60
A	352.84	341.68	557.50	U	37.14	94.72	179.60
A	309.78	371.71	538.50	U	22.86	84.46	244.80
A	432.24	357.35	393.70	U	33.29	116.76	180.10
L	142.80	120.34	210.30	U	9.57	42.96	197.31
L	305.74	199.97	222.45	U	7.79	25.75	171.29
L	309.67	164.40	206.32	U	6.07	36.85	174.20
L	325.76	151.63	201.90	U	108.14	96.16	180.45
L	256.18	145.33	189.20	U	24.79	109.86	209.70
L	242.42	196.08	187.10	C	15.22	83.35	177.40
L	373.26	166.62	249.70	C	265.84	116.69	188.70
L	382.45	222.31	219.96	C	385.13	118.58	191.70
L	228.30	181.83	368.40	C	634.93	164.80	232.56
L	14.02	55.52	245.90	C	519.88	397.32	220.10
L	445.39	455.62	286.67	C	844.45	154.68	175.10
L	300.05	469.89	287.40	C	10.22	83.98	180.44
L	1133.39	581.08	613.60	C	194.83	228.07	293.60
L	652.03	517.47	410.78				

Exercise 64. In the webpage (www.wiley.com/go/glahn/practical), you can find a file named "Hydrochem.csv," containing an extensive data set of the geochemical composition of water samples from several rivers and tributaries of the Llobregat river, the most important river in the Barcelona province (NE Spain). This data was studied in detail by Otero et al. (2005) and Tolosana-Delgado et al. (2005), and placed, with the authors consent, in the public domain within the R package "compositions" (Boogaart and Tolosana-Delgado 2008). Table 8.8 provides a random sample for illustration.

Fit an ANOVA model to these three-part compositions. Draw the data set using colors to distinguish between the four rivers. Plot the means of the four groups, as estimated by the ANOVA fit.

Advanced exercise:

Extract the variance–covariance matrix of each group. Draw *confidence* regions for each of their means, as explained in Sections 7.4 and A.2. Obtain the pooled variance–covariance matrix (see Section 7.3), and draw its ellipse three times, around the centers of the three groups. Compare the different ellipses and the different matrices. Is the assumption of equal variances between the three groups reasonable?

Exercise 65. Find a reclassification between Northern and Southern countries from Example 8.5 using only animal products: red meat, white meat, eggs, milk, and fish.

Exercise 66. A typical way of assessing the goodness of an LDA is to reclassify the sample assigning to each observation the class with its highest predicted probability. Among other fundamental problems, this conventional procedure can be criticized because it reduces a rich information on assignment probabilities to just one label: misclassifying an observation of group A into group B when the predicted probabilities of these two group were $((p_A, p_B) = (0.48, 0.52))$ seems less grievous than if the probabilities were $((p_A, p_B) = (0.08, 0.92))$. Given that these probabilities are indeed compositions themselves, they can be summarized in a compositionally compliant way instead of just by counting how often each category was given a larger probability. Following Example 8.5, have a look at the probabilities of assignment of Table 8.6. Fill in the classical reclassification table

	Reclassified group		
True group	East	West	total
East			
West			

An alternative compositional view to this table would be to obtain the geometric center of the vectors of probabilities within the two groups defined by the true country and arrange the resulting numbers in a square table. Upscale these means to sum up to the number of observations within each class. Compare the two reclassification tables.

Exercise 67. Between three groups, three borders can be drawn (A with B, B with C, A with C). Show that these three boundaries intersect in one single point (a *triple junction*). Find the equation of that point. Now assume that the discriminating composition has three components: note that in this case, the boundaries could be drawn as segments from the triple junction along the directions of some vectors $\mathbf{v}_{jk}^{\perp}$ orthogonal to $\mathbf{v}_{jk}$. Obtain the equations of these boundaries.

Exercise 68. Using the data set from Exercise 64, obtain the discriminant functions between rivers A, U, and L (remove C for this exercise). This may be easily done by computing the ilr coordinates in CoDaPack and exporting them to your favorite statistical software. Draw the data in the ilr plane and in a ternary diagram, using colors to distinguish between rivers. Add the group centers and the boundaries between groups. If you use R, LDA is available with function "lda" in the package "MASS."

9

Compositional processes

Compositions can evolve depending on an external parameter such as space, time, temperature, pressure, or global economic conditions. The external parameter may be continuous or discrete. In general, the evolution is expressed as a function $\mathbf{x}(t)$, where t represents the external variable and the image is a composition in $\mathcal{S}^D$. In order to model compositional processes, the study of simple models appearing in practice is very important. This allows the analyst to distinguish at a glance features in the data useful to identify the type of process governing the phenomenon. Frequently, its evolution is presented as the evolution of D components corresponding to amounts or masses, which are typically positive. Denote these vectors of amounts by $\mathbf{z}(t)$, with values in $\mathbb{R}^D_+$, which, after closure, give the compositional process $\mathbf{x}(t) = C[\mathbf{z}(t)]$. Comparing visually the curves of evolution of $\mathbf{z}(t)$ with $\mathbf{x}(t)$ may be confusing, and inferring the behavior of $\mathbf{z}(t)$ from the behavior of $\mathbf{x}(t)$ may be quite hard or even impossible. This will be made evident in the following examples. Moreover, an apparently complicated behavior when represented in a ternary diagram may be close to a linear process in the geometry of the simplex. The main challenge is usually to identify compositional processes from available data. This is done using a variety of techniques that depend on the data, the selected model of the process, and the prior knowledge about them.

The next sections present three simple examples of such processes. A simple but important model is a linear process in the simplex; it follows a straight line in the Aitchison geometry of the simplex. Mixtures are also important, but they are not so simple when dealt with in this geometry, as they are nonlinear processes. A third type of process is the so-called *settling* process. Settling processes are studied through some idealized examples that contribute to understand the nature of

Modeling and Analysis of Compositional Data, First Edition.
Vera Pawlowsky-Glahn, Juan José Egozcue and Raimon Tolosana-Delgado.
© 2015 John Wiley & Sons, Ltd. Published 2015 by John Wiley & Sons, Ltd.

compositional evolution. This chapter also introduces compositional derivatives and some elementary differential equations in the simplex. Differential equations are a powerful tool to produce interpretable models of evolution. Many processes in the simplex can be obtained from simple differential equations that help understanding compositional evolution.

There is a huge amount of literature on the statistical fitting of models to observed evolution data. Most of these methods can be applied to compositional evolution after expressing compositions in coordinates, but they are out of the scope of this book. Some techniques of fitting linear differential models to data have been studied in Brewer et al. (2008).

9.1 Linear processes

A linear process is characterized by the fact that its trajectory is a straight line in the simplex or, if preferred, that its coordinates describe a straight line in the real space. Other characterizations are possible, for instance, in terms of a simplicial differential equation, which is addressed in Section 9.5. When observed in absolute abundances or masses, not in proportions, they frequently appear as exponential growth or decay of the components. Also frequent is the representation as logistic curves, depending on the evolution of the observed total amount corresponding to the composition. See Section 9.5 for further discussion. The growth of a population of bacteria gives a good example of this kind of processes.

Consider D different species of bacteria that proliferate in a rich medium, and assume there are no interactions between species. Under these conditions, it is well known that the absolute abundances $z_i(t)$ of bacteria species i grows proportionally to its previous amount,

$$\frac{dz_i}{dt} = \lambda_i z_i,$$

where t is time and d/dt denotes the time derivative. This causes an exponential growth of the absolute abundance of each species. Each component of the vector $\mathbf{z}(t) = [z_1, z_2, \ldots, z_D](t)$ represents the absolute abundance of one species at time t, and the model is

$$\mathbf{z}(t) = \mathbf{z}(0) \cdot \exp(\lambda t), \tag{9.1}$$

where $\lambda = [\lambda_1, \lambda_2, \ldots, \lambda_D]$ contains the rates of growth corresponding to the species. In this case, λ_i will be positive, but one can imagine $\lambda_i = 0$, that is, the ith species does not vary; or $\lambda_i < 0$, that is, the ith species abundance decreases with time. Model (9.1) represents a process in which both the total absolute abundance of bacteria and the composition of the abundances by species (relative abundances) are specified.

Frequently, actual measurements of the bacteria population are proportions between the species and, accordingly, they do not provide information on the total abundance. In this situation, interest is centered in the compositional aspect of Equation (9.1), which is readily obtained applying a closure to it, that is, $x(t) = C[z(t)]$, taking values in S^D, and rewriting it using the operations of the simplex. This leads to

$$x(t) = x(0) \oplus t \odot p, \qquad p = \exp(\lambda), \tag{9.2}$$

where a straight line in the simplex is identified: $x(0)$ is a point on the line taken as the origin; p is a constant vector representing the direction of the line; and t is a parameter with values on the real line (positive or negative).

The vector p, being a composition, is scale invariant and can be thus multiplied by any positive constant without modifying the resulting line. This is the same as adding any constant to the vector of rates. That is, $\lambda' = \lambda + \lambda_0 1$ is equivalent to λ for any real λ_0. Thus, the signs of each individual λ_i are arbitrary and cannot be interpreted as individual growth or decay of each species, as in the model for $z(t)$. Actually, in the compositional model, only the differences between any two pairs of rates $\lambda_i - \lambda_j$ are meaningful, and they can be interpreted as relative evolution rates of species i against species j. In a compositional framework, a canonical choice of vector λ' is to force it to sum up to zero, that is, $\lambda_0 = (\sum \lambda_i)/D$, which makes $\lambda' = \text{clr}(p)$.

The linear character of Equation (9.2) is enhanced when it is represented using coordinates. Select a basis in S^D, for instance, using balances determined by a sequential binary partition, and denote the coordinates $u(t) = \text{ilr}(x)(t)$, $v = \text{ilr}(p)$. The model for the coordinates is then

$$u(t) = u(0) + t \cdot v, \tag{9.3}$$

a typical expression of a straight line in $\mathbb{R}^{D-1}$. The processes that follow a straight line in the simplex are more general than those represented by Equations (9.2) and (9.3), because changing the parameter t by any function $\phi(t)$ in the expression (Equation 9.3) still produces images on the same straight line. Section 9.5.1 presents an alternative way of obtaining this compositional result by employing a differential equation.

Example 9.1 (Growth of bacteria population).
Consider three species ($D = 3$), whose relative abundances were 82.7%, 16.5%, and 0.8% at the initial observation ($t = 0$). The rates of growth are known to be $\lambda_1 = 1$, $\lambda_2 = 2$, and $\lambda_3 = 3$. Select the sequential binary partition and balances specified in Table 9.1. The process of growth is shown in Figure 9.1, both in a ternary diagram (Figure 9.1a) and in the plane of the selected coordinates

Table 9.1   Sequential binary partition and balance
coordinates used in Example of 9.1.

Order	x_1	x_2	x_3	Balance coordinates
1	+1	+1	−1	$u_1 = \dfrac{1}{\sqrt{6}} \ln \dfrac{x_1 x_2}{x_3^2}$
2	+1	−1	0	$u_2 = \dfrac{1}{\sqrt{2}} \ln \dfrac{x_1}{x_2}$

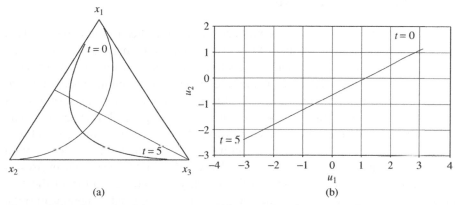

(a)   (b)

*Figure 9.1   Growth of three species of bacteria in five units of time. (a) Ternary
diagram; axis used are shown (thin lines). (b) Process in coordinates.*

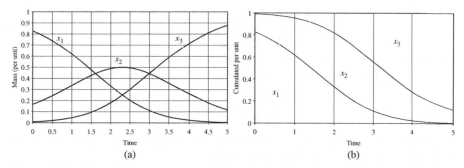

(a)   (b)

*Figure 9.2   Growth of three species of bacteria in five units of time. Evolution
of relative abundance for each species. (a) Relative abundance. (b) Cumulative
relative abundance adding to 1; x_1, lower band; x_2, intermediate band; x_3 upper
band. Note the inversion of abundances of species 1 and 3.*

(Figure 9.1b). Using coordinates, it is easy to identify that the process corresponds to a straight line in the simplex. Figure 9.2 shows the evolution of the process in time in two usual plots: Figure 9.2a shows the evolution of each component in *per unit*; in Figure 9.2b, the same evolution is presented as parts adding up to one in a cumulative form. Normally, the graph in Figure 9.2a is easier to understand from the point of view of evolution, although it should be interpreted with caution: one might be tempted to say that species x_2 changes its behavior around $t = 2.25$ (from growth to decay), although its slope is actually constant. ◊

9.2   Mixture processes

The action of mixing two or more matters can be simplified to a tank model. This assumes that there are a number of reservoirs containing materials characterized by their compositions. The matter in each tank is then transferred to a new container at a certain rate. Then, the material coming from different tanks is agitated to get a homogeneous material, which is the mixed matter. There are several ways of mixing materials. Some of them are linear but, in general, they are not. The following example is a typical mixing process that is nonlinear in the simplex.

Consider two tanks partially filled with D species of materials or liquids with mass (or volume) concentrations given by $\mathbf{x}$ and $\mathbf{y}$ in S^D. The total masses in the tanks are m_1 and m_2, respectively. Initially, the concentration in the first tank is $\mathbf{z}_0 = \mathbf{x}$. The content of the second container is steadily poured into the first one, and this is stirred. The mass transferred from the second tank to the first one is $m_2\phi(t)$ at time t, that is, $\phi(t)$ is the proportion of mass in the second container already poured into the first container. The evolution of mass in the first container is

$$(m_1 + \phi(t)m_2) \cdot \mathbf{z}(t) = m_1 \cdot \mathbf{x} + \phi(t)m_2 \cdot \mathbf{y},$$

where $\mathbf{z}(t)$ is the process of the concentration in the first container. Note that $\mathbf{x}$, $\mathbf{y}$, $\mathbf{z}$ are considered closed to 1. The final composition in the first container is

$$\mathbf{z}_1 = \frac{1}{m_1 + m_2}(m_1\mathbf{x} + m_2\mathbf{y}). \tag{9.4}$$

The mixture process can be alternatively expressed as mixture of the initial and final compositions (often called end points):

$$\mathbf{z}(t) = \alpha(t)\mathbf{z}_0 + (1 - \alpha(t))\mathbf{z}_1,$$

for some function of time, $\alpha(t)$, where, to fit the physical statement of the process, $0 \leq \alpha \leq 1$. But there is no problem in assuming that α may take values on the whole real line.

Example 9.2 (Obtaining a mixture).
A container A contains a mixture of three liquids. The number of volume units in A for each component are $[30, 50, 13]$, that is, the composition in ppu (parts per unit) is $\mathbf{z}_0 = \mathbf{z}(0) = [0.3226, 0.5376, 0.1398]$. Another mixture of the three liquids, $\mathbf{y}$, is in container B. The content of B is poured and stirred in A. The final concentration in A is $\mathbf{z}_1 = [0.0411, 0.2740, 0.6849]$. One can ask for the composition $\mathbf{y}$ and for the required volume in container B. Using the notation introduced above, the initial volume in A is $m_1 = 93$, the volume and concentration in B are unknown. Equation (9.4) is now a system of three equations with three unknowns: m_2, y_1, y_2 (the closure condition implies $y_3 = 1 - y_1 - y_2$):

$$
m_1 \begin{pmatrix} z_1 - x_1 \\ z_2 - x_2 \\ z_3 - x_3 \end{pmatrix} = m_2 \begin{pmatrix} y_1 - z_1 \\ y_2 - z_2 \\ 1 - y_2 - y_3 - z_3 \end{pmatrix}, \tag{9.5}
$$

which, being a simple system, is not linear in the unknowns. Note that Equation (9.5) involves masses or volumes and, therefore, it is not a purely compositional equation. This situation always occurs in mixture processes. Figure 9.3 shows the process of mixing (M) both in a ternary diagram (Figure 9.3a) and in the balance coordinates $u_1 = 6^{-1/2} \ln(z_1 z_2 / z_3)$, $u_2 = 2^{-1/2} \ln(z_1/z_2)$ (Figure 9.3b). Figure 9.3 also shows a simplicial linear process, that is, a straight line in the simplex, going from $\mathbf{z}_0$ to $\mathbf{z}_1$ (P).

It should be noted that the fact that this mixing process (M) is nonlinear in the simplex, or that the process P is linear in the simplex, is not related to the speed transfer of mass between B and A, or of the composition $\mathbf{z}(t)$ going from $\mathbf{z}_0$ to $\mathbf{z}_1$.   ◇

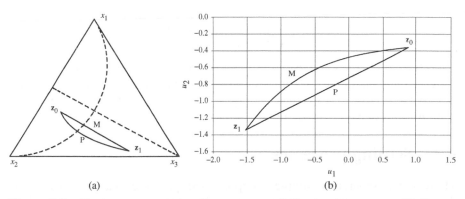

(a)   (b)

Figure 9.3   Two processes going from $\mathbf{z}_0$ to $\mathbf{z}_1$. (M) mixture process; (P) linear perturbation process. Representation in the ternary diagram (a) and using balance coordinates $u_1 = 6^{-1/2} \ln(z_1 z_2 / z_3)$, $u_2 = 2^{-1/2} \ln(z_1/z_2)$ (b).

The next example describes a mixing process that is linear in the simplex because of the fact that the concentrations of the mixed species remain constant along the process.

Example 9.3 (washing process).
A reservoir of constant volume V receives an input flow Q (volume per unit time) and, after a very active mixing inside the reservoir, an equivalent flow Q is released. At time $t = 0$, masses $m_1(0)$, $m_2(0)$, $m_3(0)$ of three contaminants are stirred and homogeneously dispersed in the reservoir. After stirring, the concentrations of the contaminants will be $\mathbf{x}(0) = C[m_1(0), m_2(0), m_3(0)]$. The contaminant species are assumed nonreactive. Attention is paid to the relative content of the three contaminant species at the output in time. The output concentration of a contaminant is proportional to the mass in the reservoir (Albarède, 1995, p. 346),

$$m_i(t) = m_i(0) \cdot \exp\left(-\frac{t}{V/Q}\right), \quad i = 1, 2, 3,$$

which represents an exponential decay of the masses of the contaminants. After joining these three equations in a composition and applying the closure, this process corresponds to a linear process in S^3 as shown in

$$\mathbf{x}(t) = C[m_1(t), m_2(t), m_3(t)] = \mathbf{x}(0) \oplus t \odot \exp\left[-\frac{Q}{V}, -\frac{Q}{V}, -\frac{Q}{V}\right].$$

The peculiarity is that, in this case, $\lambda_i = -Q/V$ for the three species. A representation in orthogonal coordinates, as functions of time, is

$$u_1(t) = \sqrt{\frac{2}{3}} \ln \frac{(x_1(t)x_2(t))^{1/2}}{x_3(t)} = \sqrt{\frac{2}{3}} \ln \frac{(x_1(0)x_2(0))^{1/2}}{x_3(0)},$$

$$u_2(t) = \frac{1}{\sqrt{2}} \ln \frac{x_1(t)}{x_2(t)} = \frac{1}{\sqrt{2}} \ln \frac{x_1(0)}{x_2(0)}.$$

Therefore, from the compositional point of view, the relative concentration of the contaminants in the subcomposition associated with the three contaminants is constant. This is not contradictory with the fact that the total mass of contaminants decays exponentially in time.   ◇

9.3   Settling processes

Many other apparently simple compositional processes happen to be non-linear in the simplex. This is the case of systems in which the mass from some components is transferred into other ones, possibly, but not necessarily, preserving the total mass.

For example, consider masses of R radioactive isotopes $\{m_1, m_2, \ldots, m_R\}$ that disintegrate into nonradioactive materials $\{m_{R+1}, m_{R+2}, \ldots, m_D\}$. The process in time t is described by

$$m_i(t) = m_i(0) \cdot \exp(-\lambda_i t), \quad m_j(t) = m_j(0) + \sum_{i=1}^{R} a_{ij}(m_i(0) - m_i(t)),$$

for $1 \leq i \leq R$ and $R + 1 \leq j \leq D$. The positive coefficients a_{ij} control how decomposed materials are distributed on the various components m_j, and they satisfy that $\sum_{i=1}^{R} a_{ij} \leq 1$. When this sum is equal to 1, the mass of the system is constant; otherwise some mass is lost. From the compositional point of view, the interest is centered in the composition $\mathbf{x}(t) = C\mathbf{m}(t)$. The subcomposition corresponding to the first group of parts ($1 \leq i \leq R$) behaves as a linear process. The second group of parts $\{x_{R+1}, x_{R+2}, \ldots, x_D\}$ is called *settled group* because it receives what has been lost from other parts, possibly preserving the total mass in the system. This kind of settling processes does not evolve linearly in the simplex, despite its simple form.

Example 9.4 (One radioactive isotope).
Consider the radioactive isotope m_1 that is transformed into the nonradioactive isotope m_3, while the element m_2 remains stable. This situation, with $\lambda_1 < 0$, corresponds to

$$m_1(t) = m_1(0) \cdot \exp(\lambda_1 t), \quad m_2(t) = m_2(0), \quad m_3(t) = m_3(0) + m_1(0) - m_1(t),$$

which is mass-preserving. The compositional process is $\mathbf{x}(t) = C\mathbf{m}(t)$. The group of parts behaving linearly is $\{x_1, x_2\}$, and the settling group is $\{x_3\}$.

Table 9.2   Parameters for Example 9.4: *one radioactive isotope*. Disintegration rate is $\ln 2$ times the inverse of the half-lifetime. Time units are arbitrary. The lower part of the table represents the sequential binary partition used to define the balance coordinates.

Parameter	x_1	x_2	x_3
Disintegration rate	0.5	0.0	0.0
Initial mass	1.0	0.4	0.5
Balance 1	+1	+1	−1
Balance 2	+1	−1	0

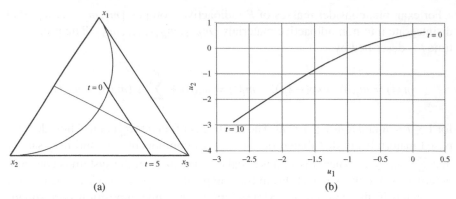

*Figure 9.4   Disintegration of one isotope x_1 into x_3 in 10 units of time.
(a) Ternary diagram; axis used are shown (thin lines). (b) Process in
coordinates. The mass in the system is constant and the mass of x_2 is constant.*

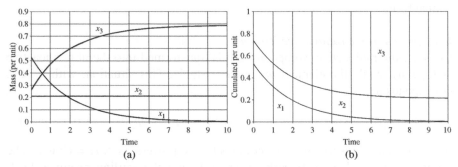

*Figure 9.5   Disintegration of one isotope x_1 into x_3 in 10 units of time.
Evolution of Proportions of mass for each species. (a) Proportions of mass.
(b) Cumulated Proportions of mass; x_1, lower band; x_2, intermediate band; x_3
upper band. Note the inversion of abundances of species 1 and 3.*

Table 9.2 shows parameters of the model, and Figures 9.4 and 9.5 show different
aspects of the compositional process from $t = 0$ to $t = 10$.

  A first inspection of the figures reveals that the process appears as a segment in
the ternary diagram (Figure 9.4b). This fact is due to the constant mass of x_2 in
a conservative system, thus appearing as a constant per unit. In Figure 9.4a, the
evolution of the coordinates shows that the process is not linear; however, except
for initial times, the process may be approximated by a linear one. The linear or
nonlinear character of the process is hardly detected in Figures 9.5 showing the
evolution in time of the composition.   ◇

Example 9.5 (Three radioactive isotopes).
Consider three radioactive isotopes that we identify with a group of parts, $\{m_1, m_2, m_3\}$, evolving linearly in the simplex. The disintegrated mass of m_1 is distributed on the nonradioactive parts $\{m_4, m_5, m_6\}$ (settling group). The whole disintegrated mass from m_2 and m_3 is assigned to m_5 and m_6, respectively. The corresponding composition is $\mathbf{x}(t) = C\mathbf{m}(t)$. The values of the parameters

Table 9.3   Parameters for Example 9.5: *three radioactive isotopes.*
Disintegration rate is ln 2 times the inverse of the half-lifetime. Time units are arbitrary. The middle part of the table corresponds to the coefficients a_{ij} indicating the part of the mass from the x_i component transformed into x_j. Note that they add to one and the system is mass conservative. The lower part of the table shows the sequential binary partition to define the balance coordinates.

Parameter	x_1	x_2	x_3	x_4	x_5	x_6
Disintegration rate	0.2	0.04	0.4	0.0	0.0	0.0
Initial mass	30.0	50.0	13.0	1.0	1.2	0.7
Mass from x_1	0.0	0.0	0.0	0.7	0.2	0.1
Mass from x_2	0.0	0.0	0.0	0.0	1.0	0.0
Mass from x_3	0.0	0.0	0.0	0.0	0.0	1.0
Balance 1	+1	+1	+1	−1	−1	−1
Balance 2	+1	+1	−1	0	0	0
Balance 3	+1	−1	0	0	0	0
Balance 4	0	0	0	+1	+1	−1
Balance 5	0	0	0	+1	−1	0

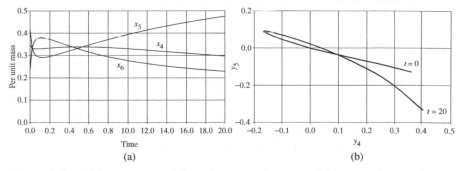

Figure 9.6   Disintegration of three isotopes x_1, x_2, x_3. Disintegration products are masses added to x_4, x_5, x_6 in 20 units of time. (a) Evolution of per unit of mass of x_4, x_5, x_6. (b) x_4, x_5, x_6 process in coordinates; a loop and a double point are revealed.

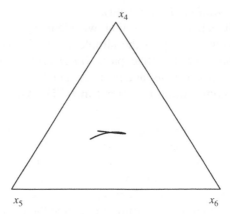

Figure 9.7   Disintegration of three isotopes x_1, x_2, x_3. Products are masses added to x_4, x_5, x_6, in 20 units of time, represented in the ternary diagram. Loop and double point are visible.

considered are shown in Table 9.3. Figure 9.6a shows the evolution of the subcomposition of the settling group in 20 time units; no special conclusion is obtained from it. Contrarily, Figure 9.6b, showing the evolution of the coordinates of the subcomposition, reveals a loop in the evolution with a double point (the process passes two times through this compositional point); although less clearly, the same fact can be observed in the representation in the ternary diagram in Figure 9.7. This is a quite surprising and complex behavior despite the very simple character of the settling process. Changing the parameters of the process, one can obtain a simpler behavior, for instance, without double points or exhibiting less curvature. However, these processes only present one possible double point or a single bend point; the branches far from these points are suitable for a linear approximation.   ◇

Example 9.6 (Washing process revisited).
Consider the washing process in Example 9.3. Let us assume that the liquid is water with density equal to 1 and define the mass of water $m_0(t) = V \cdot 1 - \sum m_i(t)$. This may be considered as a settling process. The mass concentration at the output of the reservoir acts as the closure of the four components. The compositional process is not a straight line in the simplex, because a new balance is needed to represent the process. This balance can be

$$y_0(t) = \sqrt{\frac{3}{4}} \ln \frac{(x_1(t)x_2(t)x_3(t))^{1/3}}{x_0(t)},$$

which is neither a constant nor a linear function of t.   ◇

9.4   Simplicial derivative

The derivative plays an important role in the analysis of real functions, as it describes the rate of change of a function when the argument of the function increases. Similarly, the change of a composition, as a function of an argument, can be described using derivatives. As attention is focused on evolution of compositions in time, the composition is assumed to be a function of an argument denoted as t, which ranges a real interval $T \subseteq \mathbb{R}$. However, t can represent any real variable, such as space along a line, pressure, temperature, indices of activity (real), and so on. Then, simplex-valued (sv) functions $\mathbf{x} : T \subseteq \mathbb{R} \to S^D$ are now considered as models of compositional processes. In general sv-functions are denoted as $\mathbf{x}(t)$ or with other bold letters.

In real calculus, the derivative of a real function $f : T \subseteq \mathbb{R} \to \mathbb{R}$ is a real function $\partial f(t)$ defined as

$$\partial f(t) = \lim_{h \to 0} \frac{1}{h}(f(t+h) - f(t)), \tag{9.6}$$

if the limit exists. Note that the standard notation $df(t)/dt$ has been replaced by ∂. By analogy to the definition of derivative of a real function, the simplicial derivative of an sv-function $\mathbf{x}(t)$ is

$$\partial_\oplus \mathbf{x}(t) = \lim_{h \to 0} \frac{1}{h} \odot (\mathbf{x}(t+h) \ominus \mathbf{x}(t)), \tag{9.7}$$

provided that the limit exists. It is remarkable that $\partial_\oplus \mathbf{x}(t)$ is obtained changing the real operations in $\partial f(t)$ by their homologous operations in the simplex; but operations in the simplex are applied to compositions that are vectors and not to real-single-valued functions. When defined for t in an interval T, the simplicial derivative $\partial_\oplus \mathbf{x}(t)$ is again an sv-function.

Simplicial derivatives are mentioned in Aitchison et al. (2002) or in Aitchison (1986, Postscript to the 2003 reprint). A review of properties of simplicial derivative is found in Egozcue et al. (2011) or Egozcue and Jarauta-Bragulat (2014).

The definition of simplicial derivative in Equation (9.7) is useful for interpretation, but it is not appropriate for computation. Following the principle of working on coordinates, an alternative expression using ilr-coordinates is readily obtained:

$$\partial_\oplus \mathbf{x}(t) = \mathrm{ilr}^{-1}[\partial(\mathrm{ilr}(\mathbf{x}(t)))], \tag{9.8}$$

where the ordinary derivative operator ∂ applies componentwise. Similar expressions hold using clr or even just logarithms:

$$\partial_\oplus \mathbf{x}(t) = \mathrm{clr}^{-1}[\partial(\mathrm{clr}(\mathbf{x}(t)))] = C\left[\exp(\partial \ln[\mathbf{x}(t)])\right]. \tag{9.9}$$

The last expression was used in the earlier definitions of simplicial derivative (Aitchison et al., 2002). Equations (9.8) and (9.9) can be formally applied to nonclosed vectors. If $\mathbf{y}(t)$ is an $\mathbb{R}_+^D$-function, such that $C\left[\mathbf{y}(t)\right] = \mathbf{x}(t)$, the computation is still $\partial_\oplus \mathbf{x}(t) = C\left[\exp(\partial \ln[\mathbf{y}(t)])\right]$. This means that closure plays a secondary role in the computation, recalling that $\mathbf{x}(t)$ is a representative of the equivalence class of $\mathbf{y}(t)$.

Property 9.7 Some elementary simplicial derivatives are as follow:

1. Consider $\mathbf{x}(t), \mathbf{y}(t)$ sv-functions and α, β real constants; then the simplicial derivative is linear in the simplex:

$$\partial_\oplus[(\alpha \odot \mathbf{x}) \oplus (\beta \odot \mathbf{y})] = (\alpha \odot \partial_\oplus \mathbf{x}) \oplus (\beta \odot \partial_\oplus \mathbf{y}).$$

2. For $\mathbf{x}(t) = \mathbf{u} \in S^D$ a constant composition, the derivative is $\partial_\oplus \mathbf{u} = \mathbf{n}$, where $\mathbf{n}$ is the neutral element of S^D that, remember, plays the role of the zero vector.

3. The derivative of a monomial sv-function $\mathbf{x}(t) = t^k \odot \mathbf{z}$, where $\mathbf{z}$ is a constant composition, is

$$\partial_\oplus(t^k \odot \mathbf{z}) = kt^{k-1} \odot \mathbf{z},$$

and a polynomial sv-function has derivative

$$\partial_\oplus \left(\bigoplus_{k=0}^{m} t^k \odot \mathbf{z} \right) = \bigoplus_{k=1}^{m} kt^{k-1} \odot \mathbf{z}.$$

4. The case $k = 1$ is important in itself, as it states that the derivative of a linear process (Section 9.1) is constant:

$$\partial_\oplus(\mathbf{a} \oplus t \odot \mathbf{z}) = \mathbf{z}, \tag{9.10}$$

where $\mathbf{a}$ is an arbitrary constant composition.

Details are given in Egozcue et al. (2011).

The first-order approximation of an sv-function $\mathbf{x}(t)$ at a point t_0 provides the concepts of local linearization in the simplex and the tangent sv-function at this point. The tangent sv-function to $\mathbf{x}(t)$ at t_0 is the linear sv-function

$$\mathbf{r}(t) = \mathbf{x}(t_0) \oplus [(t - t_0) \odot \partial_\oplus \mathbf{x}(t_0)],$$

which, whenever the derivative exists, locally approximates $\mathbf{x}(t)$ in a neighborhood of t_0. Conceptually, $\mathbf{r}(t)$ is the first-order Taylor polynomial and gives an interpretation of the simplicial derivative as a rate of change of $\mathbf{x}(t)$.

Example 9.8 (Tangent approximation to a mixing process).
A three-part mixing process is described by the sv-function

$$\mathbf{x}(t) = C[u_1 + t(v_1 - u_1), u_2 + t(v_2 - u_2), u_3 + t(v_3 - u_3)], \quad 0 \le t \le 1,$$

where $\mathbf{u} = [u_1, u_2, u_3]$, $\mathbf{v} = [v_1, v_2, v_3]$ are constant compositions in $\mathcal{S}^3$. For instance, a process $\mathbf{x}(t)$ starts at $t = 0$ at the composition $\mathbf{u} = [0.900, 0.001, 0.099]$ and moves to $\mathbf{x}(1) = \mathbf{v} = [0.700, 0.299, 0.001]$. This kind of mixing process was described in Section 9.2 and its nonlinearity in the simplex was demonstrated. Using Equation (9.9), the simplicial derivative of $\mathbf{x}(t)$ is

$$\partial_\oplus \mathbf{x}(t) = C \exp \left[\frac{v_1 - u_1}{u_1 + t(v_1 - u_1)}, \frac{v_2 - u_2}{u_2 + t(v_2 - u_2)}, \frac{v_3 - u_3}{u_3 + t(v_3 - u_3)} \right].$$

Figure 9.8 shows the evolution of the process (dotted and dashed curves) both in proportions (Figure 9.8a) and in coordinates (Figure 9.8b). Coordinates were taken as two balances defined as

$$b_1 = \sqrt{\frac{2}{3}} \ln \frac{x_1}{(x_2 x_3)^{1/2}}, \quad b_2 = \frac{1}{\sqrt{2}} \ln \frac{x_2}{x_3}. \tag{9.11}$$

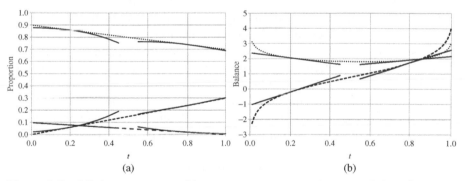

Figure 9.8   Mixing process and its tangent process at times $t = 0.2$ and $t = 0.8$. (a) Representation in proportions. (b) Representation in balances. Components of the process are represented as dotted and dashed lines; components of tangent processes as solid lines.

The tangent processes at $t = 0.2$ and $t = 0.8$ have been also plotted (solid lines) both in proportions and in coordinates. The parts of the mixing process in proportions appear as straight lines, while the tangent processes are curved. This appearance is reversed when represented in coordinates. It can be observed that the tangent processes are quite good local approximations near to the tangent points. ◇

The concept of simplicial derivative is easily extended to higher order derivatives. Provided that $\partial_{\oplus}\mathbf{x}(t)$ is an sv-function, the derivative operator can be applied again if $\partial_{\oplus}\mathbf{x}(t)$ is derivable in the interval of interest. The resulting simplicial derivative

$$\partial_{\oplus}^2\mathbf{x}(t) = \partial_{\oplus}[\partial_{\oplus}\mathbf{x}(t)],$$

is called second-order simplicial derivative and, if it exists is again an sv-function. The concept is easily extended to kth order derivatives that are denoted by $\partial_{\oplus}^k\mathbf{x}(t)$.

9.5   Elementary differential equations

Differential equations involve unknown functions and their derivatives and, if they have solution(s), they can be considered as models of evolution in time. When the unknown function is an sv-function and the derivative is simplicial, they are called simplicial differential equations. Herein, simplicial differential equations are viewed as a way of obtaining models of compositional processes with special emphasis in processes that come from simple equations. Some of those models are also simple when formulated directly on the sv-function itself, but there are cases in which, although the sv-function is quite complex, the corresponding simplicial differential equation is easily interpretable.

A simplicial differential equation only gives relative information between the parts involved, and it does not convey any information about absolute abundances or masses. This fact, although obvious from the principles of compositional analysis, may cause misinterpretations and surprise to the analyst used to interpret models in terms of absolute abundances.

First-order differential equations can be expressed as implicit expressions, which are out of the scope of this introduction. Attention is paid to explicit first-order simplicial differential equations, that is, where the derivative is isolated,

$$\partial_{\oplus}\mathbf{x}(t) = \mathbf{f}(\mathbf{x}(t), t), \tag{9.12}$$

with $\mathbf{f} : S^D \times T \rightarrow S^D$. When $\mathbf{f}(\mathbf{x}(t), t)$ does not depend on t explicitly, that is, $\mathbf{f}(\mathbf{x}(t), t)$ is reduced to $\mathbf{f}(\mathbf{x}(t))$, the simplicial differential equation is called *autonomous*. This case is important for modeling processes in the simplex. If $\mathbf{x}(t)$ is a model of evolution of a system and obeys an autonomous differential equation, $\mathbf{x}(t)$ describes the behavior of an isolated system free of forcing external actions, thus offering the opportunity of modeling the internal interactions between the parts of the evolving composition.

After stating and interpreting a simplicial differential equation, it is important to seek for methods of finding its solutions. The main tool consists of translating it into ilr-coordinates. For instance, taking ilr-transformation on the left-hand side of Equation (9.12) yields

$$\text{ilr}(\partial_\oplus \mathbf{x}(t)) = \partial[\text{ilr}(\mathbf{x})(t)] = \partial \mathbf{x}^*(t),$$

where $\mathbf{x}^*(t)$ is the vector of $D-1$ coordinates of $\mathbf{x}(t)$. The ilr-transformed right-hand side of Equation (9.12) can be written as

$$\text{ilr}[\mathbf{f}(\mathbf{x}(t), t)] = \text{ilr}[\mathbf{f}(\text{ilr}^{-1}[\text{ilr}(\mathbf{x}(t))], t)] = \mathbf{f}^*(\mathbf{x}^*(t), t),$$

where the last term is the vector of $D-1$ coordinates of $\mathbf{f}$, which arguments are the coordinates $\mathbf{x}^*(t)$ and t. Equating both sides, a system of $D-1$ ordinary differential equations (ODEs) is obtained, that is,

$$\partial \mathbf{x}^*(t) = \mathbf{f}^*(\mathbf{x}^*(t), t). \tag{9.13}$$

Such a system of equations (9.13) can be solved using the standard techniques, thus transforming the problem of solving a differential equation in the simplex (Equation 9.12) into a system of ODEs (9.13). Note that the system (Equation 9.13) has a different expression for different ilr-transformations, but the solutions of Equation (9.13) are different coordinate expressions of the same sv-function $\mathbf{x}(t)$. Therefore, solution of Equation (9.12) does not depend on the specific ilr-coordinates used for solving the equation.

9.5.1   Constant derivative

The simplest simplicial differential equation is $\partial_\oplus \mathbf{x}(t) = \mathbf{z}$, where $\mathbf{z} \in S^D$ and $\mathbf{x}(t)$ is an sv-function on S^D representing a compositional process. When $\mathbf{z} = \mathbf{n}$, the obvious solution is $\mathbf{x}(t) = \mathbf{a}$, with $\mathbf{a}$ a constant composition in S^D: it does not reflect any change in the initial composition and, therefore, is a trivial constant process. If $\mathbf{z} \neq \mathbf{n}$, Equation (9.10) implies that the solution is $\mathbf{x}(t) = \mathbf{a} \oplus (t \odot \mathbf{z})$, that is, the process $\mathbf{x}(t)$ is represented as a straight line in the simplex. As extensively commented before, the process $\mathbf{x}(t) = \mathbf{a} \oplus (t \odot \mathbf{z})$ can be identified as the compositional evolution underlying an exponential growth or decay of masses or abundances. In fact, the D positive components of $\mathbf{z}$ can be expressed as $z_i = \exp(\lambda_i)$ or, equivalently, as $\lambda_i = \ln z_i$, $i = 1, 2, \ldots, n$. Denoting $\lambda = [\lambda_1, \lambda_2, \ldots, \lambda_n]$, the equation and its solution are

$$\partial_\oplus \mathbf{x}(t) = C \exp(\lambda),$$

$$\mathbf{x}(t) = \mathbf{a} \oplus t \odot C \exp(\lambda) = \mathbf{a} \oplus C \exp[\lambda_1 t, \lambda_2 t, \ldots, \lambda_n t].$$

However, the compositional process $\mathbf{x}(t)$ could be obtained from nonexponential evolutions of the absolute masses or abundances of the components.

An immediate alternative is to make explicit the closure in Equation (9.14). A simple but striking example of this fact follows.

Example 9.9 (The XIX-th century Malthus-Verhulst controversy).
In his celebrated contributions (1832–1839), Malthus (Malthus, 1990; Bacaër, 2011) proposed that populations grow exponentially in time, starting a vivid controversy about the future of the human population. One of the most qualified counteropinions was defended by Verhulst (1838) who proposed the logistic curve or sigmoid as a model of population growth. Despite this controversy, both models correspond to the same equation when viewed from the compositional viewpoint.

To state the compositional problem, assume that at time t, there is a quantity $m(t)$ of available resources. These can be divided into two parts: consumed resources by the present population, denoted by $y_1(t)$, and $y_2(t)$ remaining nonconsumed resources, with $y_1(t) + y_2(t) = m(t)$. The component $y_1(t)$ is taken as a proxy of the population. Assume that the compositional process $\mathbf{x}(t) = C[y_1(t), y_2(t)]$ is a solution of the simplicial differential equation $\partial_\oplus \mathbf{x}(t) = \mathbf{z}$, with $\mathbf{z} = C[\exp(\lambda_1), \exp(\lambda_2)]$. As in Equation (9.14), the compositional solution can be expressed as

$$\mathbf{x}(t) = \left[\frac{a_1 \exp(\lambda_1 t)}{a_1 \exp(\lambda_1 t) + a_2 \exp(\lambda_2 t)}, \quad \frac{a_2 \exp(\lambda_2 t)}{a_1 \exp(\lambda_1 t) + a_2 \exp(\lambda_2 t)} \right],$$

for arbitrary positive constants a_1, a_2. The resource process is then $\mathbf{y}(t) = m(t)\mathbf{x}(t)$ whichever is the evolution of the positive mass $m(t)$. The proportions $a_1/(a_1 + a_2)$, $a_2/(a_1 + a_2)$ are readily identified as the initial conditions $x_1(0), x_2(0)$, respectively.

The Malthus proposal is that $y_1(t) = m(t)x_1(t) = m(0)x_1(0)\exp(\lambda_1 t)$. This expression is obtained from the general solution taking $\lambda_2 = 0$ and $m(t) = m(0)(a_1 \exp(\lambda_1 t) + a_2)$. In this case, the second component is constant, $y_2(t) = m(0)a_2$. Hence, the Malthusian model corresponds to the composition satisfying $\partial_\oplus \mathbf{x}(t) = C[\exp(\lambda_1), 1]$, where the remaining resources are constant in time.

The Verhulst logistic curve is obtained setting $\lambda_2 = 0$ and $m(t) = m(0)$, that is, the total resources are constant in time and the population (consumed resources) is

$$y_1(t) = \frac{m(0)a_1 \exp(\lambda_1 t)}{a_2 + a_1 \exp(\lambda_1 t)},$$

a traditional logistic curve or sigmoid.

As a conclusion, both models of consumed and remaining resources correspond to the same simplicial differential equation. They only differ in the assumed total

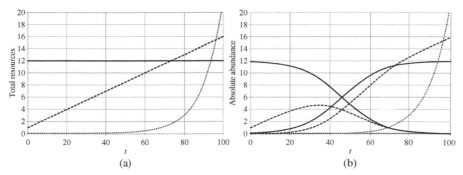

Figure 9.9   Three cases of resource evolution (consumed, remaining) with common underlying compositional process in Equation (9.14). Solid lines: constant total resources (Verhulst); dotted lines: constant remaining resources (Malthus); dashed lines: linear increase of total resources. (a) Evolution of total resources m(t). (b) Paired curves of consumed resources, $y_1(t)$, $y_2(t)$; consumed resources are always increasing with t. In the three cases, $\lambda_1 = 0.1$ and $\lambda_2 = 0$.

resources considered: Malthus proposed that remaining resources are constant in time, while Verhulst hypothesizes that what is constant is the total resources.

Figure 9.9b shows consumed and remaining resources under three different hypotheses on resources: constant total resources (Verhulst), constant remaining resources (Malthus), and linearly increasing total resources. Figure 9.9a shows the evolution of total resources in the three cases. The consumed and remaining resource curves (Figure 9.9a) appear to be substantially different. However, the proportions of consumed and remaining resources change in time exactly in the same way. These proportions can be visualized, up to the closing constant, as the solid lines in Figure 9.9b.   ◇

9.5.2   Forced derivative

Interest is now in the simplicial differential equation

$$\partial_\oplus \mathbf{x}(t) = \mathbf{f}(t), \tag{9.14}$$

where $\mathbf{f}(t)$ is an sv-function. It represents a system that tends to evolve linearly, as described in Section 9.5.1, when the system is not externally influenced. However, the sv-function $\mathbf{f}(t)$ can be interpreted as an external influence that diverts the system from its linear autonomous evolution. The form of $\mathbf{f}(t)$ is very general, and its expression should be selected according to the system and external actions to be modeled.

For many $\mathbf{f}(t)$, the solutions of the differential equations are explicit. To obtain these solutions, the expression in coordinates is useful. Taking ilr-coordinates in Equation (9.14), the system of $D - 1$ ODEs is

$$\partial \mathbf{x}^*(t) = \mathbf{f}^*(t), \tag{9.15}$$

where $\mathbf{x}^*(t)$ and $\mathbf{f}^*(t)$ are the ilr-coordinates of $\mathbf{x}(t)$ and $\mathbf{f}(t)$, respectively. The system (Equation 9.15) is easily solved because of two characteristics: (i) the system of $D - 1$ differential equations is not coupled, that is, each equation does not depend on other equations; and (ii) each equation is directly integrable. Assuming that initial conditions at $t = 0$ are given, $\mathbf{x}^*(0) = \mathbf{x}_0^*$, the solution is

$$\mathbf{x}^*(t) = \mathbf{x}_0^* + \int_0^t \mathbf{f}^*(\tau)\, d\tau. \tag{9.16}$$

The corresponding solution in S^D is readily obtained applying ilr^{-1}.

The case $\mathbf{f}(t) = \mathbf{z}$, where $\mathbf{z} \in S^D$ is constant, has been studied in Section 9.5.1. Two cases of $\mathbf{f}(t)$ are illustrated: exponential polynomials and exponential sinusoidals.

Consider the case in which the components of $\mathbf{f}(t)$ are exponential polynomials $f_i(t) = C \exp(a_{0i} + a_{1i}t + \ldots + a_{ki}t^k)$. The coordinates of $\mathbf{f}^*(t)$ are polynomials in t and their degrees are less than or equal to k. Hence, the integral in Equation (9.16) is easily carried out. The following example shows the potential applications of the solutions of such equations as models of compositional evolution.

Example 9.10 (Survival of bacteria).
Consider a culture of bacteria containing three different species not interacting with each other. At time $t = 0$, the proportions of these species are known to be $\mathbf{x}(0) \in S^3$. At that instant of time, a drug is inoculated in the culture. It is known that all bacteria will be killed by the drug, but the effect is delayed differently on each species. For each species, the hazard function $\lambda_i(t) = a_i + b_i t$, $\lambda_i \geq 0$ for $t \geq 0$, is known. The question is which is the evolution of the proportions of the three species after the inoculation of the drug.

In survival analysis, a hazard function for the ith species is introduced as the probability of one individual (in this case, each individual bacterium) to die in the time interval $[t, t + dt)$ conditional to its survival before t. If T_i is the random time of surviving for one bacteria of the ith species, and dt is a short period of time, the hazard function satisfies

$$\lambda_i(t)\, dt \simeq \Pr[t \leq T_i < t + dt \mid T_i > t] = \frac{F_i(t + dt) - F_i(t)}{1 - F_i(t)},$$

where $F_i(t)$ is the cumulative distribution function (cdf) of time until death T_i. Taking the limit $dt \to 0$, the expression of the hazard function is

$$\lambda_i(t) = \frac{\partial F_i(t)}{1 - F_i(t)} = -\partial \ln(1 - F_i(t)) = -\partial \ln(S_i(t)),$$

where $S_i(t) = 1 - F_i(t)$ is known as the survival function of the ith species. The survival functions are obtained from the hazard functions as

$$S_i(t) = \exp\left(-\int_0^t \lambda_i(\tau)\, d\tau\right) = \exp\left[-a_i t - \frac{1}{2}b_i t^2\right].$$

The proportions of the three species considered at time t are proportional to the product $x_i(0) \cdot S_i(t)$. Grouped in a composition, they are $\mathbf{x}(t) = \mathbf{x}_0 \oplus \mathbf{S}(t)$, where $\mathbf{S}(t) = C[S_1(t), S_2(t), S_3(t)]$. At this point, the desired process $\mathbf{x}(t)$ has been found, because all involved parameters are known. However, it is interesting to make explicit the corresponding simplicial differential equation. Taking simplicial derivative, it yields

$$\partial_\oplus \mathbf{x}(t) = \partial_\oplus \mathbf{S}(t) = C \exp[\partial \log \mathbf{S}(t)] = C \exp[-\lambda(t)] = C \exp[-\mathbf{a} - \mathbf{b}t],$$

where $\mathbf{a} = [a_1, a_2, a_3]$ and $\mathbf{b} = [b_1, b_2, b_3]$. This simplicial differential equation can be transformed into coordinates. Using the balances given in Equation (9.11) as coordinates, the system of two ODEs is

$$\partial x_1^*(t) = \sqrt{\frac{2}{3}} \left[(-a_1 - b_1 t) + \frac{1}{2}[(a_2 + b_2 t) + (a_3 + b_3 t)]\right],$$

$$\partial x_2^*(t) = \frac{1}{\sqrt{2}} \left[(-a_2 - b_2 t) + (a_3 + b_3 t)\right],$$

which, after grouping terms of equal power of t, gives

$$\partial x_1^*(t) = \sqrt{\frac{2}{3}} \left[\frac{a_2 + a_3 - 2a_1}{2} + \frac{b_2 + b_3 - 2b_1}{2}t\right],$$

$$\partial x_2^*(t) = \frac{1}{\sqrt{2}} \left[(a_3 - a_2) + \frac{b_3 - b_2}{2}t\right],$$

where the derivatives of the coordinates are equated to polynomials of first degree.

For illustration, two cases are shown in Figure 9.10. The chosen parameters are given in Table 9.4. Both cases have equal parameters except in the survival function of the first species S_1. In the first case (shown in Figure 9.10, left panels), species 1 is unaltered by the drug but its possible growth is inhibited. Thus, it

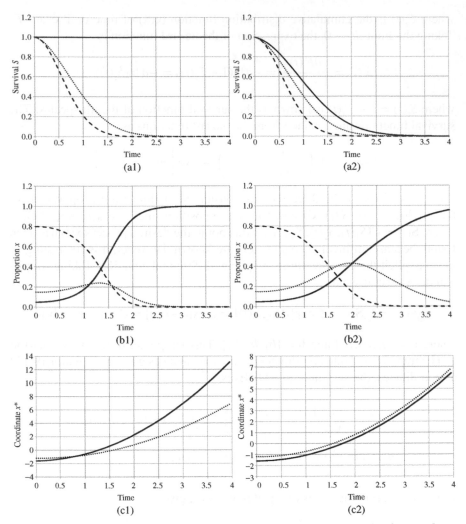

Figure 9.10   Survival of three bacteria species in two cases. (a1, a2) panels: survival functions. (b1, b2) panels: evolution of proportions of the three species. (c1, c2) panels: evolution of coordinates.

appears as a reference, as its individuals neither die nor reproduce. In the second case, the first species is also affected by the drug but is more resistant than the other two species, as shown in Figure 9.10, panels (a1) and (a2). The solutions of the corresponding differential equations in proportions are shown in panels (b1) and (b2). The fact that species 1 dominates the other species in about 2 units of

Table 9.4   Parameters of survival example in two cases. The initial proportions are $x_i(0)$. The survival functions are specified by the parameters a_i and b_i. Parameters of x_1 are given for cases 1 and 2. For case 1, species x_1 is stationary, that is, it neither grows nor decays.

	x_1 (case 1)	x_1 (case 2)	x_2	x_3
$x_i(0)$	0.05	0.05	0.15	0.80,
a_i	0.00	0.01	0.15	0.05
b_i	0.00	1.00	1.50	3.00

time is not surprising in the first case, as this species is not affected by the drug. Remarkably, this also occurs in the second case, because of the longer survival time of species 1. It should be taken into account that the evolution within the composition $[x_1(t), x_2(t), x_3(t)]$ is represented in proportions and not in absolute abundances. Under the assumption of the first case, the absolute abundance of species 1 would be constant in time, whereas in the second case, an exponential decrease of the three species is expected. The result is that panels (b1) and (b2) are qualitatively similar. Panels (c1) and (c2) show the evolution of the coordinates $x_1^*(t)$ and $x_2^*(t)$, where the similarity between the two cases clearly appears again.   ◇

In many cases, the dynamics of a composition can be affected by a periodic forcing term. Diurnal and annual cycles are typical examples in geoenvironmental cases. The corresponding models based on simplicial differential equations can have $\mathbf{f}(t)$ forcing terms in which components are exponential sinusoidals,

$$\mathbf{f}(t) = C \exp[a_1 \cos(\omega t + \phi_1), a_2 \cos(\omega t + \phi_2), \dots, a_n \cos(\omega t + \phi_n)], \quad (9.17)$$

where ω represents the angular frequency, the amplitudes $a_i \geq 0$ control the intensity of the effect on the corresponding component of the derivative, and ϕ_i accounts for the phase of the effect. For instance, some aerosols naturally decay in time, but the solar radiation accelerates or dampens the decay. Equation (9.14), with forcing terms of the form (9.17) or similar ones, can be simple models for this kind of phenomena. Figure 9.11a shows the evolution of proportions for such an equation with period equal to 5 time units, $n = 3$, $a_1 = 0.5$, $a_2 = 1$, $a_3 = 3$, $\phi_1 = 0$, $\phi_2 = \pi/4$, $\phi_3 = \pi/3$. Initial conditions were selected as $\mathbf{x}(0) = [0.1, 0.4, 0.5]$. Figure 9.11b shows the behavior of the coordinates, which are taken as in Equation (9.11).

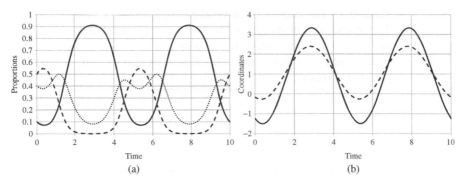

Figure 9.11   Solution of Equation (9.11). (a) Proportions ($x_1(t)$, full line; $x_2(t)$, dotted line; $x_3(t)$, dashed line). (b) Coordinates as defined in Equation (9.11) ($x_1^*(t)$ full line; $x_2^*(t)$ dashed line).

9.5.3   Complete first-order linear equation

Constant coefficients, linear, first-order, simplicial differential equations (SLODE) provide a very flexible and manageable family of models matching the behavior of many phenomena, as happens with constant coefficient, linear, first-order real differential equations. The general expression of these equations is

$$\partial_\oplus \mathbf{x}(t) = \mathbf{x}(t) \boxdot \mathbf{A} \oplus \mathbf{f}(t), \ t \in T \subseteq \mathbb{R}, \tag{9.18}$$

where $\mathbf{A}$ is a matrix of a linear transformation of the simplex (endomorphism, Section 4.9.2). The sv-function $\mathbf{f}(t)$ represents a time-dependent external action forcing the compositional process. It is commonly designed to match the known characteristics of the phenomenon to be modeled. When $\mathbf{f}(t)$ is constant in time, the differential equation in Equation (9.18) becomes an autonomous equation, that is, the equation does not explicitly depend on time. These autonomous cases are specially interesting, because the differential equation reflects all the internal interactions between the parts of the evolving composition. For this reason, this section is restricted to the autonomous case, and $\mathbf{f}(t)$ is, therefore, taken as a constant $\mathbf{f}$.

As in previous cases, Equation (9.18) can be transformed into coordinates. Let $\boldsymbol{\Psi}$ be the contrast matrix associated with a particular choice of the ilr-coordinates. Taking ilr-transformation of Equation (9.18), the differential equation of the coordinates becomes

$$\partial \mathbf{x}^*(t) = \mathbf{x}^*(t) \cdot \mathbf{A}^* + \mathbf{f}^*, \tag{9.19}$$

where $\mathbf{x}^*(t) = \mathrm{ilr}(\mathbf{x})$ and ∂ is the ordinary derivative with respect to t and the endomorphism $\mathbf{A}^* = \boldsymbol{\Psi} \mathbf{A} \boldsymbol{\Psi}^\top$ as given in Section 4.9.2. Equation (9.19) is a system of

$D - 1$ ODEs with constant coefficients – the entries of matrix $\mathbf{A}^*$. This system of differential equations can be explicitly solved using algebraic methods (e.g., Boyce and DiPrima, 2009). However, the qualitative characteristics of the solutions depend on the eigenvalues and eigenvectors of the endomorphism represented by matrix $\mathbf{A}^*$: these eigenvalues and eigenvectors are intrinsic characteristics of the endomorphism and they do not depend on the basis chosen to solve the equation. In particular, the $D - 1$ non-null eigenvalues of $\mathbf{A}$ are the same as the eigenvalues of $\mathbf{A}^*$ in *any* orthonormal basis, independently of the matrix $\boldsymbol{\Psi}$.

Stability is one of these system characteristics. A point $\mathbf{x}_c^*$ in $\mathbb{R}^{D-1}$ is called a critical point if $\partial \mathbf{x}_c^* = 0$; for Equation (9.19), the only critical point is $\mathbf{x}_c^* = -(\mathbf{A}^*)^{-1}\mathbf{f}^*$ whenever $\mathbf{A}^*$ is invertible. The main characteristic of critical points is their stability. Roughly speaking, a critical point is stable if, for large values of t, the solutions of the equations remain at a bounded distance from the critical point. Particularly, the critical point is asymptotically stable if the solutions converge to the critical point. The critical point $\mathbf{x}_c^*$ is stable whenever all the eigenvalues of $\mathbf{A}^*$ are nonnegative (e.g., Pontriaguin, 1969).

Stability of the critical point of the coordinate equation (9.19) is extremely important to predict the behavior of the solutions of the corresponding simplicial equation (9.18). In fact, an unbounded unstable solution of a coordinate, necessarily corresponds to a solution in the simplex for which some of the parts approach either zero or one, where the infinity of the simplex is placed. An asymptotically stable solution approaches the value $\mathbf{x}_c = \mathrm{ilr}^{-1}(\mathbf{x}_c^*) \in S^D$ as t increases; a stable, but not asymptotically stable, solution in the simplex is maintained in a neighborhood of $\mathbf{x}_c$. Stability is then an important hint when modeling a long-term compositional phenomenon. If all parts should maintain their values near to a composition $\mathbf{x}_c$, unstable models are inappropriate. Alternatively, if one or more parts are likely to vanish with time, any stable model conceptually fails at long-term prediction.

Some examples with three or four parts are given in the following. In these cases, the ODE in Equation (9.19) is a system of two or three differential equations, respectively. For three-part equations, the critical points are then classified into nodes, spiral foci, centers, and saddle points. Nodes and spiral foci can be stable or unstable; centers are stable; and saddle points are always unstable (e.g. Lefschetz, 1977; Boyce and DiPrima, 2009).

Example 9.11 (Stable focus).
Consider the simplicial differential equation $\partial_\oplus \mathbf{x}(t) = \mathbf{x} \boxdot \mathbf{A} \oplus \mathbf{f}$ in S^3 where

$$\mathbf{A} = \begin{pmatrix} -0.5988954 & 0.2846154 & 0.3142800 \\ 0.2846154 & -0.3857200 & 0.1011046 \\ 0.3142800 & 0.1011046 & -0.4153846 \end{pmatrix}, \mathbf{f}^\mathrm{T} = \begin{pmatrix} 0.05167024 \\ 0.37609922 \\ 0.57223054 \end{pmatrix}.$$

$$\text{(9.20)}$$

Matrix **A** fulfills the *zero sum property* (Property 4.16). In order to work in coordinates, consider the balances

$$b_1 = \sqrt{\frac{2}{3}} \ln \frac{(x_1 x_2)^{1/2}}{(x_3)}, \quad b_2 = \frac{1}{\sqrt{2}} \ln \frac{x_1}{x_2} \qquad (9.21)$$

and their corresponding contrast matrix $\boldsymbol{\Psi}$. In these coordinates, **A** and **f** correspond to

$$\mathbf{A}^* = \boldsymbol{\Psi}\mathbf{A}\boldsymbol{\Psi}^\mathsf{T} = \begin{pmatrix} -0.6230769 & -0.1846154 \\ -0.1846154 & -0.7769231 \end{pmatrix},$$

$$(\mathbf{f}^*)^\mathsf{T} = (\mathrm{ilr}(\mathbf{f}))^\mathsf{T} = (-1.153036 - 1.403587).$$

The eigenvalues of $\mathbf{A}^*$ are -0.5, -0.9, which correspond to an asymptotically stable node in the critical point $\mathbf{x}_c^* = [-1.414881, -1.470387]$, which, backtransformed into proportions, is $\mathbf{x}_c = [0.04, 0.32, 0.64]$ (Figure 9.12). ◇

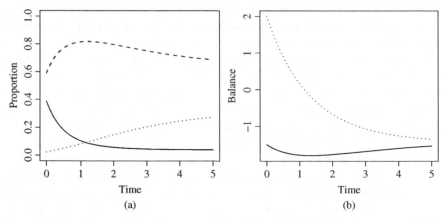

Figure 9.12   Compositional process defined by the SLODE in Equation (9.19) with coefficients shown in Equation (9.20). Evolution of (a) proportions and (b) the corresponding balances.

Example 9.12 (An epidemic event).
A new disease is introduced in a large population. At time $t = 0$, some infected and contagious (Icont) individuals are introduced into the population. The types of individuals are susceptible-healthy (Susc), infected and contagious (Icont), infected but not contagious (Inf), and recovered and resistant (Recov).

Initially, the proportions are $\mathbf{x}(0) = [0.976, 0.0191, 0.00464, 6.11 \cdot 10^{-07}]$, a large proportion of healthy but susceptible individuals and a nonnegligible proportion of infected-contagious individuals. There is a very low proportion of resistant individuals. The epidemic event is assumed to follow a process modeled by the SLODE in Equation (9.18). The matrix of the system and its constant term are shown in Table 9.5. The eigenvalues of the matrix $\mathbf{A}^*$ are -0.018, -0.028, and 0.030. As the eigenvalues of $\mathbf{A}^*$ are real, the matrix is symmetric and $\mathbf{A}$ is also symmetric (see Table 9.5). As the third eigenvalue is positive, the process is not stable, that is, some proportions tend to zero for large times. Figure 9.13 shows the evolution of the outbreak both in proportions (Figure 9.13a) and balance coordinates (Figure 9.13b), as defined in Table 9.6. Proportions show a quite common shape in these types of events: a small proportion of individuals (Icont) are able to infect almost the whole population in a short period of time. In this case, there is a proportion of infected individuals who are not able to transmit the disease (Inf), and they are finally converted into recovered-resistant individuals (Recov). Also contagious-infected individuals (Icont) are able to recover from the disease. This evolution of proportions seems quite complicated, but the evolution of the balances reveals a simple and almost linear behavior (Figure 9.13b). As expected, the process is unstable and some balances tend in time to infinity and, accordingly, the recovered (Recov) proportion tends to one,

Table 9.5   Coefficients of the system in Example 9.12. The left four rows correspond to the matrix $\mathbf{A}$, and the fifth row to the constant term composition $\mathbf{f}$. The labels are placed to make the interpretation easy.

	$\partial_\oplus \mathbf{x}_{Susc}$	$\partial_\oplus \mathbf{x}_{Icont}$	$\partial_\oplus \mathbf{x}_{Inf}$	$\partial_\oplus \mathbf{x}_{Recov}$
$\mathbf{x}_{Susc}$	0.2171829	0.3158277	1.1460868	−1.6790973
$\mathbf{x}_{Icont}$	0.3158277	−0.3317696	−0.5968584	0.6128003
$\mathbf{x}_{Inf}$	1.1460868	−0.5968584	−0.4288066	−0.1204218
$\mathbf{x}_{Recov}$	−1.6790973	0.6128003	−0.1204218	1.1867189
$\mathbf{f}$	0.03	0.15	0.24	0.64

Table 9.6   SBP code defining coordinates in the epidemic event example

	Susc	Icont	Inf	Recov
b_1	1	−1	−1	1
b_2	1	0	0	−1
b_3	0	1	−1	0

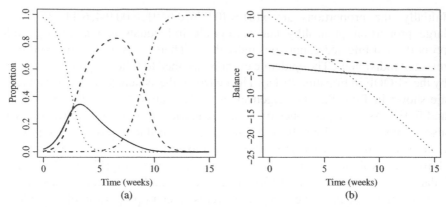

*Figure 9.13   Evolution of a fictitious epidemic event along 15 weeks.
(a) Proportions in a population. Dotted line, susceptible (healthy) individuals;
solid line, infected and contagious individuals; dashed line, infected but not
contagious individuals; dotted-dashed line, recovered and resistant individuals.
(b) Balances following the SBP coded in Table 9.6, b_1 (solid line), b_2 (dotted
line), b_3 (dashed line).*

while the others go to zero. The repetition of the event will depend on whether
a recovered individual turns into a susceptible one with time or not. Matrix
A expresses the interactions between the proportions of classes of individuals
and the components of the simplicial derivatives. The interpretation must take
into account that the entries of **A** add up to zero by rows, that is, we should
tend to understand them as coefficients of some clr-transformed object. One
obvious interaction is visualized in the fourth column labeled $\partial_\oplus x_{Recov}$, which
interpretation can be phrased as *the presence of a high proportion of susceptible
population delays the growth of recovered* (coefficient -1.6790973). Also, in a
moderate sense (entry -0.1204218), *a high proportion of infected (Inf) delays
the increase of recovered (Recov)*. Note that the equivalent reasoning can be
performed on the positive entries of this column $\partial_\oplus x_{Icont}$. The second column
of Table 9.5, labeled $\partial_\oplus x_{Icont}$, is interesting, as it is the basis on which the
populations recover from the epidemic event. Attending to the two negative
entries of the column, the conclusion is *both infected-contagious and simply
infected proportions push the proportion of contagious infected to decrease*.
That is, as the population is progressively infected, it is harder to increase the
proportion of contagious infected. Note that interpreting the positive entries of
this column is more difficult, although they might be taken as positive feedback
effects.   ◊

Example 9.13 (Stable spiral).

The balances in Equation (9.21) are considered again for a three-part simplex. They are assumed to be solutions of the system of ODEs

$$\partial \mathbf{x}^*(t) = \mathbf{x}^*(t)\mathbf{A}^* + \mathbf{f}^*, \tag{9.22}$$

where

$$\mathbf{A}^* = \begin{pmatrix} 0.15 & -1.55 \\ 5.30 & -1.10 \end{pmatrix}, \quad (\mathbf{f}^*)^{\mathsf{T}} = \begin{pmatrix} -1.465107 \\ 1.899547 \end{pmatrix}. \tag{9.23}$$

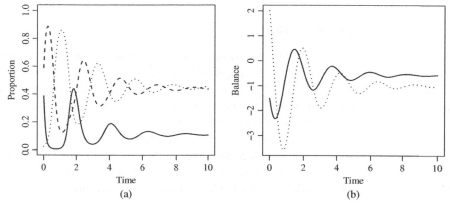

Figure 9.14   Compositional process defined by the ODE in Equation (9.22) with coefficients shown in Equation (9.23). Evolution of (a) proportions and (b) the corresponding balances.

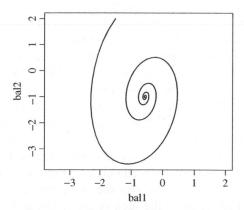

Figure 9.15   Compositional process defined by the ODE in Equation (9.22) with coefficients shown in Equation (9.23). Evolution of balances in the phase plane.

The eigenvalues of this matrix are the complex conjugate pair $-0.95 \pm$ $i\ 2.797208$. As the eigenvalues have negative real part and are complex, the equation has spiral stable solutions converging to the critical point $[-0.5659523, -1]$ in coordinates, which approximately corresponds to the composition $[0.109, 0.449, 0.442]$ (Figure 9.14). The oscillatory character of the solutions corresponds to a spiral point. Figure 9.15 shows the joint behavior of the two coordinates (phase plane) x_1^*, x_2^* converging to the critical point. The matrix $A = \Psi^{\top} A^* \Psi$ in the corresponding SLODE is not symmetric in this case, in contrast with the matrices of the two previous examples. ◊

9.5.4   Harmonic oscillator

Many natural systems oscillate as a result of combinations of opposite actions that are shifted in time or, in more physical terms, opposite forces that are not in phase. A spring is a typical physical example: when compressed, a force tries to compensate the action; when elongated, a force opposite to the previous one pushes the spring toward its equilibrium position. This kind of behavior is intuitively extended to more general systems in which compositional magnitudes play the major role. For instance, the predator–prey Lotka–Volterra equations (see, e.g., Odum, 1971; Volterra, 1926) model an ecosystem of two species, such as hares and lynxes. The predator–prey Lotka–Volterra equations are

$$\partial m_1(t) = -a_1 m_1(t) + a_2 m_1(t) m_2(t)$$
$$\partial m_2(t) = b_1 m_2(t) - b_2 m_1(t) m_2(t), \tag{9.24}$$

where $m_1(t)$ is the number of predators (lynxes), $m_2(t)$ is the number of potential preys (hares), and the total population is $m(t) = m_1(t) + m_2(t)$. The first observation on Equation (9.24) is that it is not scale invariant: if the number of individuals is expressed in thousands of individuals, the coefficients a_2 and b_2 should be multiplied by 1000 to maintain the solution of the differential equation unaltered. The conclusion is that Equation (9.24) is not a compositional model, as it also models the behavior of the total population $m(t)$. On the other hand, the rationale of Equation (9.24) is that the derivative $\partial m_1(t)$, reporting the rate of increase of predators, should decrease proportionally to their number, as they compete for territory; simultaneously, the derivative should increase proportionally to the probability of encounters between predators and preys, as indicated by the term $a_2 m_1(t) m_2(t)$. What is not clear is why the two proportional terms are joined using an addition that destroys the idea of proportionality. Similar comments hold for the equation with the second derivative, $\partial m_2(t)$. Equation (9.24) can be used to simulate the evolution of a population of hares and lynxes in an almost stationary regime. Figure 9.16 shows the behavior of a Lotka–Volterra system solution (hares–lynxes) with initial conditions $m_1(0) = 134.85$ (thousands of

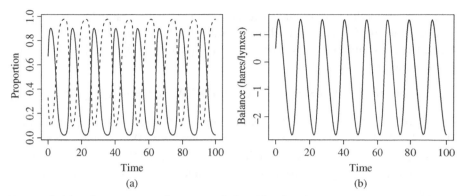

Figure 9.16   Compositional process defined by the ODE in Equation (9.22) with coefficients shown in Equation (9.23). Solution of the Lotka–Volterra equation with parameters reported in the text. (a) Evolution of proportions of hares (solid line) and lynxes (dashed line). (b) Balance hares/lynxes.

hares in 1885) and $m_2(0) = 65.69$ (thousands of lynxes in 1885) as reported in Odum (1971). The parameters adopted for the equation were

$$a_1 = -0.842, \quad a_2 = 0.019, \quad b_1 = 0.473, \quad b_2 = -0.016.$$

Figure 9.16 exhibits a clear oscillatory behavior both in proportions and in the balance. However, the balance oscillations (Figure 9.16b) differ from a pure sinusoidal curve, as the oscillation is a little bit skewed. Also proportions show a slightly different behavior around maxima or minima. This means that the solution of the Lotka–Volterra equation could be modeled by an undamped harmonic oscillator; if the model is a second-order differential equation, solutions are exactly sinusoids for the balance, but higher-order equations can approximate asymmetrical features.

In a more general setting, consider a D-part composition $\mathbf{x}(t)$ that evolves in an oscillatory way. Once a contrast matrix $\mathbf{\Psi}$ is adopted to define ilr-coordinates, the corresponding $D - 1$ coordinates $\mathbf{x}^*(t)$ represent the compositional process. As the coordinates are real variables, they can be modeled as a $(D - 1)$-dimensional second-order harmonic oscillator

$$\partial^2 \mathbf{x}^*(t) - \partial \mathbf{x}^*(t)\mathbf{W}_1^* + \mathbf{x}^*(t)\mathbf{W}_0^* = \mathbf{0}, \tag{9.25}$$

where $\mathbf{W}_1^*$, $\mathbf{W}_0^*$ are nonnegative definite $(D - 1, D - 1)$-matrices and the coordinates $\mathbf{x}^*(t)$ are gathered in a $D - 1$-row vector. Equation (9.25) admits straightforward generalizations and simplifications. The damping term related to $\partial \mathbf{x}^*(t)$ can be suppressed to obtain an undamped oscillator; a time-dependent forcing term can replace the $\mathbf{0}$-vector; and, quite importantly, the order of the leading derivative can be higher than 2. These generalizations keep the linear character of Equation (9.25).

The corresponding simplicial differential equation is obtained by taking ilr^{-1} in Equation (9.25). The second-order ordinary derivative is translated into a second-order simplicial derivative as $\partial_{\oplus}(\text{ilr}^{-1}(\partial \mathbf{x}^*(t))) = \partial_{\oplus}^2 \mathbf{x}(t)$. The ilr^{-1} of the damping term can be written as

$$\text{ilr}^{-1}(\partial \mathbf{x}^*(t)\mathbf{W}_1) = \partial_{\oplus}C\exp[(\mathbf{x}^*(t)\boldsymbol{\Psi}\boldsymbol{\Psi}^{\mathsf{T}}\mathbf{W}_1^*)\boldsymbol{\Psi}] = \partial_{\oplus}\mathbf{x}(t)\boxdot\mathbf{W}_1,$$

where the identity term $\boldsymbol{\Psi}\boldsymbol{\Psi}^{\mathsf{T}}$ has been inserted for easy operation and $\mathbf{W}_1 = \boldsymbol{\Psi}^{\mathsf{T}}\mathbf{W}_1^*\boldsymbol{\Psi}$. Similarly, the third term containing $\mathbf{W}_0^*$, which represents a restoring action, is backtransformed into $\mathbf{x}(t)\boxdot\mathbf{W}_0$ with $\mathbf{W}_0 = \boldsymbol{\Psi}^{\mathsf{T}}\mathbf{W}_0^*\boldsymbol{\Psi}$. Grouping all these terms by perturbation, the simplicial expression of the harmonic oscillator is

$$\partial_{\oplus}^2\mathbf{x}(t)\ominus\left(\partial_{\oplus}\mathbf{x}(t)\boxdot\mathbf{W}_1\right)\oplus\left(\mathbf{x}(t)\boxdot\mathbf{W}_0\right) = \mathbf{n}, \tag{9.26}$$

where $\mathbf{n}$ is the neutral element in S^D. The matrices $\mathbf{W}_1$, $\mathbf{W}_0$ are just matrix representations of two endomorphisms (Section 4.9.2), behaving in the same way as in the preceding Section and are supposed to be nonnegative definite in a standard harmonic oscillator.

Example 9.14. Consider a process $\mathbf{x}(t)$ in S^3 satisfying the simplicial differential equation (9.26), corresponding to a harmonic oscillator. This equation has been analyzed using balance coordinates

$$x_1^* = \sqrt{\frac{2}{3}}\ln\frac{(x_1x_2)^{1/2}}{x_3}, \quad x_2^* = \frac{1}{\sqrt{2}}\ln\frac{x_1}{x_2}. \tag{9.27}$$

In these coordinates, the matrices $\mathbf{W}_0^*$ and $\mathbf{W}_1^*$ have been chosen to be

$$\mathbf{W}_0^* = \begin{pmatrix} 2 & 0 \\ 0 & 1 \end{pmatrix}, \quad \mathbf{W}_1^* = \begin{pmatrix} 0.5 & -0.1 \\ 0.2 & 0.1 \end{pmatrix}, \tag{9.28}$$

which are both positive definite. These matrices translated into matrices in S^3 are, up to rounding errors,

$$\mathbf{W}_0 = \boldsymbol{\Psi}^{\mathsf{T}}\mathbf{W}_0^*\boldsymbol{\Psi} = \begin{pmatrix} 0.833 & -0.167 & -0.667 \\ -0.167 & 0.833 & -0.667 \\ -0.667 & -0.667 & 1.333 \end{pmatrix}, \tag{9.29}$$

$$\mathbf{W}_1 = \boldsymbol{\Psi}^{\mathsf{T}}\mathbf{W}_1^*\boldsymbol{\Psi} = \begin{pmatrix} 0.162 & 0.120 & -0.282 \\ -0.053 & 0.104 & -0.051 \\ -0.109 & -0.224 & 0.333 \end{pmatrix}, \tag{9.30}$$

where $\boldsymbol{\Psi}$ is the contrast matrix corresponding to the selected coordinates. Note that both $\mathbf{W}_0$, $\mathbf{W}_1$ fulfill the zero sum condition (Property 4.16). The initial conditions for coordinates have been taken as

$$x_1^*(0) = 1, \quad x_2^*(0) = 3, \quad \partial x_1^*(0) = 0.5, \quad \partial x_2^*(0) = -0.2,$$

far from the equilibrium point which is the $(0, 0)$ point in coordinates.

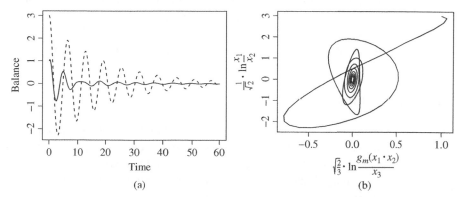

Figure 9.17   Compositional process defined by Equation (9.27) with coefficients shown in Equations (9.29) and (9.30). Solution of the harmonic oscillator in coordinates with coefficients given in Equation (9.28). (a) Evolution of the two balances: x_1^* solid line; x_2^* dashed line. (b) Evolution in the phase plane $x_1^*(t)$ versus $x_2^*(t)$.

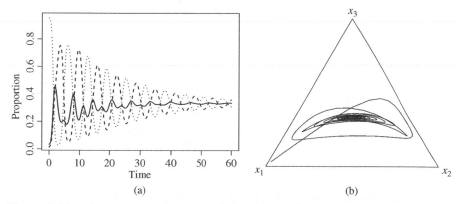

Figure 9.18   Compositional process defined by Equation (9.27) with coefficients shown in Equations (9.29), (9.30), and (9.28). (a) Evolution of proportions of the three parts. (b) Orbit in a ternary diagram.

The equation has been solved in coordinates. Figure 9.17 shows the solution obtained in coordinates. Figure 9.17b shows a quite standard figure of a damped oscillation, but some irregularities due to the nondiagonal damping matrix $\mathbf{W}_1^*$ can be observed. These minor irregularities produce the involved behavior shown in the phase plane (Figure 9.17a), where the main characteristic is the convergence of the solution to the origin. The results in Figure 9.17 can be translated to the simplex. Figure 9.18 shows the solution in time (Figure 9.18a) and the orbit in a ternary diagram (Figure 9.18b). A curious effect shows up: while proportions $x_1(t)$ (dashed curve) and $x_2(t)$ (dotted line) have a quite regular oscillation and damping, $x_3(t)$ (solid line) exhibits irregular oscillations. This was hard to foresee looking at the balances. At the same time, it is a demonstration of the flexibility of this kind of models.   ◊

9.6   Exercises

Exercise 69. Select two arbitrary three-part compositions, $\mathbf{x}(0)$, $\mathbf{x}(t_1)$, and consider the linear process from $\mathbf{x}(0)$ to $\mathbf{x}(t_1)$. Determine the direction of the process normalized to one and the time, t_1, necessary to arrive from $\mathbf{x}(0)$ to $\mathbf{x}(t_1)$. Plot the process (a) in a ternary diagram, (b) in balance coordinates, and (c) as evolution in time of the parts normalized to a constant.

Exercise 70. Consider the linear process starting with $\mathbf{x}(0) = [82.7, 16.5, 0.8]$ and rates of decay $\lambda_1 = -3$, $\lambda_2 = -2$, and $\lambda_3 = -1$. Draw the equivalents of Figures 9.1 and 9.2; for example, you can take time steps between 0 and 5 at intervals of 0.05 time units, and calculate $\mathbf{x}(t)$ for each one of the resulting 101 time moments. Use the same basis as in Example 9.1. Compare the resulting diagrams with Figures 9.1 and 9.2. What can you conclude with regard to the signs of each individual λ_i? Repeat for $\lambda = [-1, 0, +1]$, if you deem it necessary.

Exercise 71. Chose $\mathbf{x}(0)$ and $\mathbf{p}$ in S^3. Consider the process $\mathbf{x}(t) = \mathbf{x}(0) \oplus t \odot \mathbf{p}$ with $0 \leq t \leq 1$. Assume that the values of the process at $t = j/50, j = 1, 2, \ldots, 50$ are perturbed by observation errors, $\mathbf{y}(t)$ distributed as a normal on the simplex $\mathcal{N}_s(\boldsymbol{\mu}, \boldsymbol{\Sigma})$, with $\boldsymbol{\mu} = C[1, 1, 1]$ and $\boldsymbol{\Sigma} = \sigma^2 I_3$ (I_3 unit (3×3)-matrix). Observation errors are assumed independent of t and $\mathbf{x}(t)$. Plot $\mathbf{x}(t)$ and $\mathbf{z}(t) = \mathbf{x}(t) \oplus \mathbf{y}(t)$ in a ternary diagram and a balance-coordinate plot. Try with different values of σ^2.

Exercise 72. In Example 9.3, set $x_1(0) = 1$, $x_2(0) = 2$, $x_3(0) = 3$, $V = 100$, $Q = 5$. Find the sequential binary partition used in the example. Plot the evolution in time of the coordinates and mass concentrations including the water $x_0(t)$. Plot, in a ternary diagram, the evolution of the subcomposition x_0, x_1, x_2.

Exercise 73. In Example 9.2, find the necessary volume m_2 and the composition in container B, $\mathbf{y}$. Find the direction of the simplicial linear process to go from $\mathbf{z}_0$ to $\mathbf{z}_1$.

Exercise 74. A container has a constant volume $V = 100$ vu (vu = volume units) and initially contains a liquid whose composition is $\mathbf{x}(0) = C[1, 1, 1]$. A constant flow of $Q = 1$ vu per second, with volume composition $\mathbf{x} = C[80, 2, 18]$, is introduced into the container. After a complete mixing, there is an output whose flow equals Q with the volume composition $\mathbf{x}(t)$ at the time t. Model the evolution of the volumes of the three components in the container using ordinary linear differential equations and solve them (Hint: these equations are easily found in textbooks, e.g., Albarède (1995, pp. 345–350)). Are you able to plot the curve for the output composition $\mathbf{x}(t)$ in the simplex without using the solution of the differential equations? Is it a mixture?

Exercise 75. Check that the simplicial derivative can be computed using Equation (9.8).

Exercise 76. Consider the settling process $\mathbf{x}(t)$ defined by the equations

$$x_1(t) = a_1 \exp(\lambda_1 t),$$
$$x_2(t) = a_2 \exp(\lambda_2 t),$$
$$x_3(t) = u_3 + a_1(1 - \exp(\lambda_1 t)) + a_2(1 - \exp(\lambda_2 t)),$$

with $t \in T$, $T = (0, t_1)$, and assuming that $x_3(0) > 0$. Compute the simplicial derivative of $\mathbf{x}(t)$ for $t \in T$, and pay attention to the behavior of $\partial_\oplus \mathbf{x}(t)$ near to a large enough end point t_1. Plot the process $\mathbf{x}(t)$ and the tangent process at some values of t, both in proportions and in balance coordinates.

Exercise 77. Consider the two-dimensional real harmonic oscillator

$$\partial^2 \mathbf{x}^*(t) + \mathbf{A}_0^* \mathbf{x}^*(t) = 0, \quad \mathbf{A}_0^* = \begin{pmatrix} 2.0 & -0.5 \\ -0.5 & 1.0 \end{pmatrix},$$

with initial conditions $\partial \mathbf{x}^*(0) = [0.5, -0.5]$ and $\mathbf{x}^*(0) = [0.1, -0.3]$. Find the simplicial differential equation when $\mathbf{x}^*(t)$ is taken as the coordinate vector of a simplicial process $\mathbf{x}(t)$, using the coordinates given in Equation (9.27). Transform both the coordinate and simplicial equations into first-order differential equations. Solve the first-order differential equation for coordinates and find the corresponding solution in the simplex. Plot the time evolution for $t = 0$ to $t = 10$, as a function of time and in the phase plane $(x_1^*(t), x_2^*(t))$ and as a ternary diagram for $\mathbf{x}(t)$. Hint: the harmonic oscillator has explicit solution, but it is recommended to solve the equations numerically.

10

Epilogue

This book is by no means exhaustive. There are many issues related to the analysis of compositional data, which have not been included in an attempt to limit its length. Other issues, closely related to the methods presented before, are still a matter of active research. They are briefly discussed here, and some references are included to guide the interested reader to relevant literature.

Principal balances

One major issue in the analysis of compositions is the selection of an orthonormal basis. The most straightforward way to do this is to calculate principal components of clr-transformed compositions. But the resulting components usually involve all parts, making it difficult to interpret them. In an attempt to approximate principal components taking advantage of the interpretability of balances obtained through a sequential binary partition (SBP), principal balances were defined by Pawlowsky-Glahn et al. (2011). Three initial, suboptimal algorithms have been proposed. Approaches based on sparse principal components were studied in (Mert et al., 2014). Some of these algorithms are available in the R-package "compositions" (Boogaart and Tolosana-Delgado, 2013).

Robust analysis of compositional data

Robustness is a general issue of statistics, related to the presence of contaminated or atypical samples, that is, of *outliers*. Roughly speaking, an outlier is usually considered to be an observation that does not belong to the population. We prefer

Modeling and Analysis of Compositional Data, First Edition.
Vera Pawlowsky-Glahn, Juan José Egozcue and Raimon Tolosana-Delgado.
© 2015 John Wiley & Sons, Ltd. Published 2015 by John Wiley & Sons, Ltd.

to define an outlier as an observation that is atypical with respect to a given or assumed distribution. The reason is that the former definition suffers of a certain circular reasoning (to infer properties of the population one should know which data do belong to it, and to know which data do or do not belong to the population, one should know its properties). Instead of trying to identify outliers and then calculate some statistics, robust statistics are defined as functions of the data set whose values do not arbitrarily change. This is achieved by arbitrarily perturbing less than a fixed fraction of the data, called the *breakdown point*. For instance, the median of a variable is a univariate robust statistic with a breakdown point of almost 50%, because roughly half of the data can be perturbed and the median will not change. In multivariate statistics, one of the most popular robust estimators of mean vector and covariance matrix is the *minimum covariance determinant* (MCD) *estimator*, based on the association between an ellipsoid and these two statistics (see Appendix A.2). The MCD estimator looks for the ellipse in which the associated matrix has the smallest determinant and that contains 50% of the data. Mean and covariance are then calculated from this subset of the data.

Filzmoser and Hron (2008) adapted the concepts of robustness to compositional data analysis, basing the MCD calculation on a set of ilr coordinates, showing that the resulting estimators of center and variation matrix do not depend on the ilr used for the calculations. The method has been extended to principal component analysis (Filzmoser et al., 2009), factor analysis (Filzmoser et al., 2009), discriminant analysis (Filzmoser et al., 2009), and for dealing with zeros (Hron et al., 2010). A complete account of robustness in compositional data analysis can be found in Filzmoser and Hron (2011). Currently, robust MCD estimators of center and spread are implemented in the package "compositions." Many more robust methods are available in the package "robCompositions" (Templ et al., 2010).

Canonical correlation, partial least squares regression, and associated techniques

Regression with compositional data was treated in Sections 8.1 (compositional response) and 8.2 (compositional input). Regression with both compositional predictor and compositional response was presented by Tolosana-Delgado and Boogaart (2013b): it looks for the logcontrast in the input composition that best explains each possible coordinate of the response composition. All these regression techniques assume that the variables playing the explanatory role are known without error and that the response has a normally distributed error. *Total least squares* (TLS) is a related technique that evenly splits the error between response and input, and partial least squares (PLS) is a generalization that allows to decide which proportion of error is assigned to the input and which to the response. The use of TLS for compositional data was explored by Fiserova and

Hron (2010). A different technique is canonical correlation (CC), which looks for the logcontrasts within two sets of parts that show the highest correlation: numerically, CC gives the same solution as TLS. Aitchison (1986) presents a pioneering case of use of CC with compositional data.

All these methods can be applied to compositional data bearing the principle of working on coordinates in mind: take coordinates within each of the compositions, use the desired method (CC, PLS, TLS, regression) to detect the adequate directions in the spaces of coordinates, and backtransform the vector(s) of coefficients to represent results as compositions following Section 8.1 or 8.2.

Regionalized compositions

The existence of spurious spatial correlation was first mentioned in Pawlowsky (1984). A first attempt to avoid the problems can be found in (Pawlowsky, 1986, in German) and (Pawlowsky-Glahn and Olea, 2004, in English), based on the additive logratio transformation. This was extended by Tolosana-Delgado (2006) to the principle of working in coordinates. It is worth mentioning that the compositional approach was used to develop an alternative to indicator kriging (IK), called *simplicial indicator kriging* (SIK) (Tolosana-Delgado et al., 2008), solving the typical drawbacks of IK, that is, order relation violation and estimates of probabilities out of the admissible range. Complete practical accounts to the spatial analysis of compositional data can be found in Tolosana-Delgado (2011, in Spanish) or Tolosana-Delgado et al. (2011, in English).

The key point of a geostatistical analysis of spatial data is to estimate and model a *variogram*, a function that measures how does the expected variability between two samples evolve when the distance between their sampling locations is modified (see, e.g., Chilès and Delfiner (2012), where geostats for compositions is also briefly reported). As in compositional data, the variance is replaced by the variation matrix, here variation-variograms can be used to study how compositions increasingly differ as they are taken at spatial locations further and further apart. Variation-variograms were introduced by Pawlowsky-Glahn and Olea (2004) (after Pawlowsky (1984)) and explored in depth by Tolosana-Delgado and Boogaart (2013a). The typical use of a variogram model is to produce optimal interpolation maps by an interpolation technique called *cokriging*. The compositional version, presented in all the references mentioned, is a direct application of the principle of working in coordinates: both the available compositional data and the variation-variogram model are expressed in a set of coordinates, cokriging is then applied, and interpolated coordinate vectors are backtransformed to the original compositional scale. Experimental versions of these tools are available in the package "compositions" (Boogaart and Tolosana-Delgado, 2013, pp. 200–206).

Time series analysis

Compositional time series analysis following the ideas behind the principle of working in coordinates (transform–analyze–backtransform) date back to Smith and Brunsdon (1989), who based the analysis on alr-transformed data. Bergman (2008) was the first to explore the use of ilr for the same task. Finally, Barceló-Vidal et al. (2011) presented a complete account of the alternative methods of time series analysis of compositional data proposed up to that moment, as well as some theory and practical issues for working with VARIMA (vector autoregressive integrated moving average) models following concepts analogous to those explained in Chapters 8 and 9. This is again an application of the principle of working on coordinates: express the data in a coordinate system, use a multivariate time series model and transform predictions back to compositions. Spectral techniques for compositional time series have also been recently put forward by Pardo-Igúzquiza and Heredia (2011), with the same coordinate approach.

Vectors of positive components

It has been mentioned that the logratio approach bears a strong resemblance to the way positive variables are treated in a lognormal framework: a positive random variable arises from a multiplicative process, and taking logs it becomes an additive process that may be reasonably modeled with a normal distribution, thus one is invited to use conventional methods on log-transformed positive data (Aitchison and Brown, 1957). This framework has also been used for some kinds of compositional data, that is, for vectors of positive components for which subcompositional coherence is meaningful, but that are not scale invariant or that carry information on a total mass or abundance. Two ideas exist to treat them: either within a multivariate lognormal framework, that is, to apply conventional multivariate methods to the log-transformed scores, or to consider them in a T-space (Pawlowsky-Glahn et al., 2013), the product space of a simplex (conveying the relative information of the composition) times a positive quantity (related to the total size). In both cases, a certain transformation is viewed as a means to calculate some coordinates, which will then be analyzed with classical multivariate tools.

Bivariate contingency tables and two-way compositions

Bivariate contingency tables are two-way arrays registering the frequency of co-occurrence of the several possible outcomes of two categorical variables, for example, the hair color and the eye color of a person, with the first being black, brown, blonde, or ginger and the second brown, blue, or green. The 12 combinations of hair and eye color can be organized in a 4 × 3 table: if this table is used to register the frequency of each combination, it gives rise to a bivariate contingency table; if it is used to express the population relative frequency of each combination, a two-way composition is obtained. Already Aitchison (1986) suggested that two-way compositional data could be promising, although such data are still scarce nowadays: up to the authors' knowledge, only one case can be found: Eynatten and Tolosana-Delgado (2008) presented a data set of sediment composition where a mass of sediment was first split in grainsize classes and then the geochemical composition of each grainsize fraction was analyzed. Egozcue et al. (2008) proposed to consider these tables as compositions and built concepts like the independence table and the interaction table using orthogonal projections in the corresponding simplex, like those of Section 4.8. These concepts offer alternatives to fitting loglinear models for the analysis of count data, and for the treatment of two-way compositions, although these methods are still in their beginnings.

Generalization to continuous compositions

Egozcue and Díaz-Barrero (2003) were the first to realize that probability density functions, being positive functions of *constant sum* (integral) to one, can be interpreted as compositions of *infinite* number of parts. This work was further developed and expanded in a series of theoretical works (Boogaart, 2005; Egozcue et al., 2006; Boogaart et al., 2010, 2014). There, it was shown that many of the geometric properties of compositions explained in this book bring powerful insights in the Bayesian statistics framework. Beyond the theoretical aspects, this approach to density functions as continuous compositions has allowed to introduce new flexible models for univariate variables (Egozcue et al., 2013), to deal with grainsize compositions in a unified approach with *standard* compositions (Tolosana-Delgado et al., 2008), or to introduce statistical techniques in the analysis of density functions, such as principal components (Tolosana-Delgado et al., 2008), multidimensional scaling (Delicado, 2008), and analysis of variance (ANOVA) (Boogaart et al., 2010).

References

Agresti, A. (1990) *Categorial Data Analysis*, John Wiley & Sons, Inc., New York, 744 pp.

Aitchison, J. (1981) A new approach to null correlations of proportions. *Mathematical Geology*, **13** (2), 175–189.

Aitchison, J. (1982) The statistical analysis of compositional data (with discussion). *Journal of the Royal Statistical Society, Series B (Statistical Methodology)*, **44** (2), 139–177.

Aitchison, J. (1983) Principal component analysis of compositional data. *Biometrika*, **70** (1), 57–65.

Aitchison, J. (1984) Reducing the dimensionality of compositional data sets. *Mathematical Geology*, **16** (6), 617–636.

Aitchison, J. (1986) *The Statistical Analysis of Compositional Data*. Monographs on Statistics and Applied Probability, Chapman & Hall Ltd., London, (Reprinted in 2003 with additional material by The Blackburn Press), 416 p.

Aitchison, J. (1990) Relative variation diagrams for describing patterns of compositional variability. *Mathematical Geology*, **22** (4), 487–511.

Aitchison, J. (1997) The one-hour course in compositional data analysis or compositional data analysis is simple, in *Proceedings of IAMG'97 – The III Annual Conference of the International Association for Mathematical Geology*, vols. **I, II** and addendum, Barcelona (E), (ed. V. Pawlowsky-Glahn), International Center for Numerical Methods in Engineering (CIMNE), Barcelona (E), 1100, pp. 3–35.

Aitchison, J., Barceló-Vidal, C., Egozcue, J.J., and Pawlowsky-Glahn, V. (2002) A concise guide for the algebraic-geometric structure of the simplex, the sample space for compositional data analysis, in *Proceedings of IAMG'02 – The VIII Annual Conference of the International Association for Mathematical Geology*, vols. **I+II**, (eds U. Bayer, H. Burger, and W. Skala), Selbstverlag der Alfred-Wegener-Stiftung, Berlin, pp. 387–392, 1106 p.

Aitchison, J., Barceló-Vidal, C., Martín-Fernández, J.A., and Pawlowsky-Glahn, V. (2000) Logratio analysis and compositional distance. *Mathematical Geology*, **32** (3), 271–275.

Aitchison, J. and Brown, J.A.C. (1957) *The Lognormal Distribution*, Cambridge University Press, Cambridge, 176 p.

Modeling and Analysis of Compositional Data, First Edition.
Vera Pawlowsky-Glahn, Juan José Egozcue and Raimon Tolosana-Delgado.
© 2015 John Wiley & Sons, Ltd. Published 2015 by John Wiley & Sons, Ltd.

Aitchison, J. and Egozcue, J.J. (2005) Compositional data analysis: where are we and where should we be heading? *Mathematical Geology*, **37** (7), 829–850.

Aitchison, J. and Greenacre, M. (2002) Biplots for compositional data. *Journal of the Royal Statistical Society, Series C (Applied Statistics)*, **51** (4), 375–392.

Aitchison, J. and Kay, J. (2003) *Possible Solution of Some Essential Zero Problems in Compositional Data Analysis*, Universitat de Girona, ISBN: 84-8458-111-X, http://ima.udg.es /Activitats/CoDaWork2003/.

Aitchison, J., Mateu-Figueras, G., and Ng, K.W. (2004) Characterisation of distributional forms for compositional data and associated distributional tests. *Mathematical Geology*, **35** (6), 667–680.

Aitchison, J. and Shen, S.M. (1980) Logistic-normal distributions. Some properties and uses. *Biometrika*, **67** (2), 261–272.

Albarède, F. (1995) *Introduction to Geochemical Modeling*, Cambridge University Press, UK, 543 p., ISBN-10: 0521578043, ISBN-13: 978-0521578042.

Anderson, T. (1984) *An Introduction to Multivariate Statistical Analysis*, Probability and Mathematical Statistics, John Wiley & Sons, Inc., New York, 675 p.

Ash, R.B. (1972) *Real Analysis and Probability*, Academic Press, Inc., New York, 476 p.

Bacaër, N. (2011) *A Short History of Mathematical Population Dynamics*, Springer, 160 p.

Bacon-Shone, J. (2003) *Modelling Structural Zeros in Compositional Data*, Universitat de Girona, ISBN: 84-8458-111-X, http://ima.udg.es/Activitats/CoDaWork2003/.

Barceló, C., Pawlowsky, V., and Grunsky, E. (1994) Outliers in compositional data: a first approach, in *Papers and Extended Abstracts of IAMG'94 – The First Annual Conference of the International Association for Mathematical Geology* (ed. C.J. Chung), Mont Tremblant, Québec, pp. 21–26.

Barceló, C., Pawlowsky, V., and Grunsky, E. (1996) Some aspects of transformations of compositional data and the identification of outliers. *Mathematical Geology*, **28** (4), 501–518.

Barceló-Vidal, C., Aguilar, L., and Martín-Fernández, J.A. (2011) *Compositional VARIMA Time Series*, John Wiley & Sons, Ltd, pp. 87–103, 378 p.

Barceló-Vidal, C., Martín-Fernández, J.A., and Pawlowsky-Glahn, V. (2001) Mathematical foundations of compositional data analysis, in *Proceedings of IAMG'01 – The VII Annual Conference of the International Association for Mathematical Geology* (ed. G. Ross), Cancun (Mex), p. 20.

Berberian, S.K. (1961) *Introduction to Hilbert Space*, Oxford University Press, New York, 206 p.

Bergman, J. (2008) *Compositional Time Series: An Application*, Universitat de Girona, http://hdl.handle.net/10256/723, http://ima.udg.es/Activitats/CoDaWork2008/.

Billheimer, D., Guttorp, P., and Fagan, W. (1997) Statistical analysis and interpretation of discrete compositional data. Technical report, NRCSE technical Report 11, University of Washington, Seattle, WA, 48 p.

Billheimer, D., Guttorp, P., and Fagan, W. (2001) Statistical interpretation of species composition. *Journal of the American Statistical Association*, **96** (456), 1205–1214.

van den Boogaart, K.G. (2005) *Statistics Structured by the Aitchison Space*, Universitat de Girona, Girona (E), ISBN: 84-8458-222-1, http://ima.udg.es/Activitats/CoDaWork05/.

van den Boogaart, K.G., Egozcue, J.J., and Pawlowsky-Glahn, V. (2010) Bayes linear spaces. *SORT – Statistics and Operations Research Transactions*, **34** (2), 201–222.

van den Boogaart, K.G., Egozcue, J.J., and Pawlowsky-Glahn, V. (2014) Bayes Hilbert spaces. *Australian and New Zealand Journal of Statistics*, doi: 10.1111/anzs.12074.

van den Boogaart, K.G. and Tolosana-Delgado, R. (2005) *A Compositional Data Analysis Package for R Providing Multiple Approaches*, Universitat de Girona, Girona (E), ISBN: 84-8458-222-1, dugi-doc.udg.edu/handle/10256/617.

van den Boogaart, K.G. and Tolosana-Delgado, R. (2008) "Compositions": a unified R package to analyze compositional data. *Computers and Geosciences*, **34** (4), 320–338.

van den Boogaart, K.G. and Tolosana-Delgado, R. (2013) *Analysing Compositional Data with R*, Springer, Heidelberg, 280 p.

van den Boogaart, K.G., Tolosana-Delgado, R., and Bren, M. (2011) *The Compositional Meaning of a Detection Limit*, CIMNE, Barcelona, ISBN: 978-84-87867-76-7.

van den Boogaart, K.G., Tolosana-Delgado, R., and Konsulke, S. (2013) Chemical equilibria in compositional data, in *Mathematics of Planet Earth* (eds E. Pardo-Igúzquiza and C. Guardiola-Albert), Springer, pp. 107–110.

Box, G.E.P. and Cox, D.R. (1964) The analysis of transformations. *Journal of the Royal Statistical Society, Series B (Statistical Methodology)*, **26** (2), 211–252.

Box, G.E.P. and Tiao, G.C. (1973) *Bayesian Inference in Statistical Analysis*, Addison-Wesley, Reading, MA, 590 p.

Boyce, W.E. and DiPrima, R.C. (2009) *Elementary Differential Equations and Boundary Value Problems*, 9th edn, John Wiley & Sons, Inc., Hoboken, NJ, 796 p.

Brewer, D., Barenco, M., Callard, R., Hubank, M., and Stark, J. (2008) Fitting ordinary differential equations to short time course data. *Philosophical Transactions of the Royal Society A*, **366**, 519–544.

Buccianti, A., Mateu-Figueras, G., and Pawlowsky-Glahn, V.E. (2006) *Compositional Data Analysis in the Geosciences: From Theory to Practice*, Special Publications, vol. **264**, Geological Society, London, 212 p.

Buccianti, A. and Pawlowsky-Glahn, V. (2005) New perspectives on water chemistry and compositional data analysis. *Mathematical Geology*, **37** (7), 703–727.

Buccianti, A., Pawlowsky-Glahn, V., Barceló-Vidal, C., and Jarauta-Bragulat, E. (1999) *Visualization and Modeling of Natural Trends in Ternary Diagrams: A Geochemical Case Study*, vols. **I and II**, Tapir, Trondheim (N), pp. 139–144.

Chayes, F. (1960) On correlation between variables of constant sum. *Journal of Geophysical Research*, **65** (12), 4185–4193.

Chayes, F. (1971) *Ratio Correlation*, University of Chicago Press, Chicago, IL, 99 p.

Chilès, J.P. and Delfiner, P. (2012) *Geostatistics – Modeling Spatial Uncertainty*, Probability and Statistics, 2nd edn, John Wiley & Sons, Inc.

Chow, Y.S. and Teicher, H. (1997) *Probability Theory: Independence, Interchangeability, Martingales*, Springer Texts in Statistics, Springer-Verlag, New York, 488 p.

Coakley, J.P. and Rust, B.R. (1968) Sedimentation in an Arctic lake. *Journal of Sedimentary Petrology*, **38**, 1290–1300.

Connor, R.J. and Mosimann, J.E. (1969) Concepts of independence for proportions with a generalization of the Dirichlet distribution. *Journal of the American Statistical Association*, **64** (325), 194–206.

Daunis-i-Estadella, J., Martín-Fernández, J.A., and Palarea-Albaladejo, J. (2008) *Bayesian Tools for Count Zeros in Compositional Data*, Universitat de Girona, 8 p., http://hdl .handle.net/10256/723, http://ima.udg.es/Activitats/CoDaWork2008/.

Delicado, P. (2008) *Comparing Methods for Dimensionality Reduction When Data are Density Functions*, Universitat de Girona, http://hdl.handle.net/10256/723, http://ima.udg .es/Activitats/CoDaWork2008/.

Dishon, M. and Weiss, G.H. (1980) Small sample comparison of estimation methods for the beta distribution. *Journal of Statistical Computation and Simulation*, **11**, 1–11.

Eaton, M.L. (1983) *Multivariate Statistics. A Vector Space Approach*, John Wiley & Sons, Inc., 512 p.

Eckart, C. and Young, G. (1936) The approximation of one matrix by another of lower rank. *Psychometrika*, **1**, 211–218.

Egozcue, J.J. (2009) Reply to "On the Harker variation diagrams;..." by J. A. Cortés. *Mathematical Geosciences*, **41** (7), 829–834.

Egozcue, J.J., Barceló-Vidal, C., Martín-Fernández, J.A., Jarauta-Bragulat, E., Díaz-Barrero, J.-L., and Mateu-Figueras, G. (2011) *Elements of Simplicial Linear Algebra and Geometry*, John Wiley & Sons, Ltd, pp. 141–157, 378 p.

Egozcue, J.J., Díaz-Barrero, J., and Pawlowsky-Glahn, V. (2008) *Compositional Analysis of Bivariate Discrete Probabilities*, Universitat de Girona, http://hdl.handle.net/10256/723, http://ima.udg.es/Activitats/CoDaWork2008/.

Egozcue, J.J. and Díaz-Barrero, J.L. (2003) *Hilbert Space on Probability Density Functions with Aitchison Geometry*, Universitat de Girona, ISBN: 84-8458-111-X, http://ima .udg.es/Activitats/CoDaWork2003/.

Egozcue, J.J., Díaz-Barrero, J.L., and Pawlowsky-Glahn, V. (2006) Hilbert space of probability density functions based on Aitchison geometry. *Acta Mathematica Sinica (English Series)*, **22** (4), 1175–1182, doi: 10.1007/s10114-005-0678-2.

Egozcue, J.J. and Jarauta-Bragulat, E. (2014) Differential models for evolutionary compositions. *Mathematical Geosciences*, **46** (4), 381–410.

Egozcue, J.J., Jarauta-Bragulat, E., and Díaz-Barrero, J.-L. (2011) *Calculus of Simplex-Valued Functions*, John Wiley & Sons, Ltd, pp. 158–175. 378 p.

Egozcue, J.J., Lovell, D., and Pawlowsky-Glahn, V. (2013) *Testing Compositional Association*, Technische Universität Wien, pp. 28–36.

Egozcue, J.J. and Pawlowsky-Glahn, V. (2005) Groups of parts and their balances in compositional data analysis. *Mathematical Geology*, **37** (7), 795–828.

Egozcue, J.J. and Pawlowsky-Glahn, V. (2006) Exploring compositional data with the CoDa-Dendrogram, in *Proceedings of IAMG'06 – The XI Annual Conference of the International Association for Mathematical Geology*, Liège (B) (eds E. Pirard, A. Dassargues, and H.B. Havenith), University of Liège, Belgium, CD-ROM.

Egozcue, J.J. and Pawlowsky-Glahn, V. (2011) *Basic Concepts and Procedures*, John Wiley & Sons, Ltd, pp. 12–28, 378 p.

Egozcue, J.J., Pawlowsky-Glahn, V., Mateu-Figueras, G., and Barceló-Vidal, C. (2003) Isometric logratio transformations for compositional data analysis. *Mathematical Geology*, **35** (3), 279–300.

Egozcue, J.J., Pawlowsky-Glahn, V., Tolosana-Delgado, R., Ortego, M.I., and van den Boogaart, K.G. (2013) Bayes spaces: use of improper distributions and exponential families. *Revista de la Real Academia de Ciencias Exactas, Físicas y Naturales, Serie A, Matemáticas (RACSAM)*, **107**, 475–486, DOI: 10.1007/s13398-012-0082-6.

Egozcue, J.J., Tolosana-Delgado, R., and Ortego, M.I. (eds) (2011) *Proceedings of the 4th International Workshop on Compositional Data Analysis (2011), Sant Feliu de Guixols, Girona, Spain*, CIMNE, Barcelona, ISBN: 978-84-87867-76-7.

Eichler, J., Hron, K., Tolosana-Delgado, R., van den Boogaart, K.G., Templ, M., and Filzmoser, P. (2013) compositionsGUI: Graphical User Environment for Compositional Data Analysis. R-project. R package version 1.0.

EUROSTAT (2014) Main tables of energy production data. Eurostat web page.

Eynatten, H.V., Barceló-Vidal, C., and Pawlowsky-Glahn, V. (2003) Modelling compositional change: the example of chemical weathering of granitoid rocks. *Mathematical Geology*, **35** (3), 231–251.

Eynatten, H.V., Pawlowsky-Glahn, V., and Egozcue, J.J. (2002) Understanding perturbation on the simplex: a simple method to better visualise and interpret compositional data in ternary diagrams. *Mathematical Geology*, **34** (3), 249–257.

Eynatten, H.V. and Tolosana-Delgado, R. (2008) *A Loglinear Model of Grain Size Influence on the Geochemistry of Sediments*, Universitat de Girona, http://hdl.handle.net/10256/723, http://ima.udg.es/Activitats/CoDaWork2008/.

Fahrmeir, L. and Hamerle, A. (1984) *Multivariate Statistische Verfahren*, Walter de Gruyter, Berlin (D), 796 p.

Feller, W. (1966) *An Introduction to Probability Theory and Its Applications*, vol. **II**, John Wiley & Sons, Inc., New York, 669 p.

Filzmoser, P. and Hron, K. (2008) Outlier detection for compositional data using robust methods. *Mathematical Geosciences*, **40** (3), 233–248.

Filzmoser, P. and Hron, K. (2011) *Robust Statistical Analysis*, John Wiley & Sons, Ltd, pp. 59–72, 378 p.

Filzmoser, P., Hron, K., and Reimann, C. (2009) Principal component analysis for compositional data with outliers. *Environmetrics*, **20**, 621–632.

Filzmoser, P., Hron, K., Reimann, C., and Garrett, R. (2009) Robust factor analysis for compositional data. *Computers and Geosciences*, **35**, 1854–1861.

Filzmoser, P., Hron, K., and Templ, M. (2009) Discriminant analysis for compositional data and robust estimation. Technical Report SM-2009-3, Department of Statistics and Probability Theory, Vienna University of Technology, Austria, 27 p.

Fiserova, E. and Hron, K. (2010) Total least squares solution for compositional data using linear models. *Journal of Applied Statistics*, **37** (7), 1137–1152.

Fréchet, M. (1948) Les éléments Aléatoires de Nature Quelconque dans une Espace Distancié. *Annales de l'Institut Henri Poincaré*, **10** (4), 215–308.

Fry, J.M., Fry, T.R.L., and McLaren, K.R. (2000) Compositional data analysis and zeros in micro data. *Applied Economics*, **32** (8), 953–959.

Gabriel, K.R. (1971) The biplot – graphic display of matrices with application to principal component analysis. *Biometrika*, **58** (3), 453–467.

Galton, F. (1879) The geometric mean, in vital and social statistics. *Proceedings of the Royal Society of London*, **29**, 365–366.

Gobierno de España, Ministerio De Empleo Y Seguridad Social (2014) Principales series – empresas inscritas en la seguridad social. Official Statistics Website. http://www.empleo .gob.es/series/ (accessed 6 August 2014).

Graf, M. (2005) *Assessing the Precision of Compositional Data in a Stratified Double Stage Cluster Sample: Application to the Swiss Earnings Structure Survey*, Universitat de Girona, Girona (E), ISBN: 84-8458-222-1, dugi-doc.udg.edu/handle/10256/617.

Greenacre, M. (2010) Logratio analysis is a limiting case of correspondence analysis. *Mathematical Geosciences*, **42**, 129–134.

Greenacre, M. (2011) Measuring subcompositional incoherence. *Mathematical Geosciences*, **43** (6), 681–693.

Holm, S. (1979) A simple sequentially rejective multiple test procedure. *Scandinavian Journal of Statistics*, **6** (2), 65–70.

Hron, K., Filzmoser, P., and Templ, M.E. (eds) (2013) *Proceedings of the 5th International Workshop on Compositional Data Analysis – CoDaWork 2013, June 3-7, 2013, Vorau, Republik Österreich*, Technische Universität Wien.

Hron, K., Templ, M., and Filzmoser, P. (2010) Imputation of missing values for compositional data using classical and robust methods. *Computational Statistics and Data Analysis*, **54** (12), 3095–3107.

Ingersoll, R.V. (1978) Petrofacies and petrologic evolution of the late cretaceous fore-arc basin, northern and central California. *Journal of Geology*, **86**, 335–352.

Ingersoll, R.V. and Suczek, C.A. (1979) Petrology and provenance of Neogene sand from Nicobar and Bengal Fans, DSDP sites 211 and 218. *Journal of Sedimentary Petrology*, **49**, 1217–1228.

Jeffreys, H. (1961) *Theory of Probability*, 3rd edn (1st edn 1939), Oxford University Press, London, 470 p.

Julious, S.A. and Mullee, M.A. (1994) Confounding and Simpson's paradox. *British Medical Journal*, **309** (6967), 1480–1481.

Kocherlakota, S. and Kocherlakota, K. (2004) *Multivariate Normal Distributions*, vol. **8**, John Wiley & Sons, Inc.

Kolmogorov, A.N. and Fomin, S.V. (1957) *Elements of the Theory of Functions and Functional Analysis*, vols. **I+II**, Dover Publications, Inc., Mineola, NY, 257 p.

Krzanowski, W.J. (1988) *Principles of Multivariate Analysis: A user's perspective*, Oxford Statistical Science Series, vol. **3**, Clarendon Press, Oxford, 563 p.

Krzanowski, W.J. and Marriott, F.H.C. (1994) *Multivariate Analysis, Part 2 – Classification, Covariance Structures and Repeated Measurements*, Kendall's Library of Statistics, vol. 2, Edward Arnold, London, 280 p.

Lancaster, H.O. (1965) The Helmert matrices. *American Mathematical Monthly*, **72** (1), 4–12.

Lefschetz, S. (1977) *Differential Equations: Geometric Theory*, 2nd edn, Dover, New York, 390 p.

Lippard, S.J., Næss, A., and Sinding-Larsen, R. (eds) (1999) *Proceedings of IAMG'99 – The V Annual Conference of the International Association for Mathematical Geology*, vols. I and II, Tapir, Trondheim (N), 784 p.

Mahalanobis, P.C. (1936) On the generalised distance in statistics. *Proceedings of the National Institute of Sciences of India*, **2** (1), 49–55.

Malthus, T.R. (1990) *Ensayo sobre el principio de la población*, Spanish Reedition of Malthus Works, Aral-bolsillo, Madrid, 534 p.

Mardia, K.V., Kent, J.T., and Bibby, J.M. (1979) *Multivariate Analysis*, Academic Press, London (GB), 518 p.

Martín-Fernández, J.A. (2001) Medidas de diferencia y clasificación no paramétrica de datos composicionales. Ph.D. thesis, Universitat Politècnica de Catalunya, Barcelona (E).

Martín-Fernández, J.A., Barceló-Vidal, C., and Pawlowsky-Glahn, V. (1998) A critical approach to non-parametric classification of compositional data. in *Advances in Data Science and Classification. Proceedings of the 6th Conference of the International Federation of Classification Societies (IFCS-98), Rome, July 21-24, 1998* (eds A. Rizzi, M. Vichi, and H. Bock), Springer-Verlag, Berlin, pp. 49–56, 677 p.

Martín-Fernández, J.A., Barceló-Vidal, C., and Pawlowsky-Glahn, V. (2000) Zero replacement in compositional data sets, in *Studies in Classification, Data Analysis, and Knowledge Organization (Proceedings of the 7th Conference of the International Federation of Classification Societies (IFCS'2000), University of Namur, Namur, 11-14 July* (eds H. Kiers, J. Rasson, P. Groenen, and M. Shader), Springer-Verlag, Berlin (D), pp. 155–160, 428 p.

Martín-Fernández, J.A., Barceló-Vidal, C., and Pawlowsky-Glahn, V. (2003) Dealing with zeros and missing values in compositional data sets using nonparametric imputation. *Mathematical Geology*, **35** (3), 253–278.

Martín-Fernández, J.A., Bren, M., Barceló-Vidal, C., and Pawlowsky-Glahn, V. (1999) *A Measure of Difference for Compositional Data Based on Measures of Divergence*, vols. I and II, Tapir, Trondheim (N), pp. 211–216.

Martín-Fernández, J.A. and Daunis-i Estadella, J. (eds) (2008) *Compositional Data Analysis Workshop – CoDaWork'08, Proceedings*, Universitat de Girona, dugi-doc.udg.edu/handle/10256/618.

Martín-Fernández, J.A., Palarea, J., and Olea, R. (2011) *Dealing with Zeros*, John Wiley & Sons, Ltd, pp. 43–58. 378 p.

Martín-Fernández, J.A., Palarea-Albadalejo, J., and Gómez-García, J. (2003) *Markov Chain Montecarlo Method Applied to Rounding Zeros of Compositional Data: First Approach*, Universitat de Girona, ISBN: 84-8458-111-X, http://ima.udg.es/Activitats/CoDaWork 2003/.

Martín-Fernández, J.A. and Thió-Henestrosa, S. (2006) *Rounded Zeros: Some Practical Aspects for Compositional Data*, Special Publications, vol. **264**, Geological Society, London, pp. 191–201, 212 p.

Mateu-Figueras, G. (2003) Models de distribució sobre el símplex. Ph.D. thesis, Universitat Politècnica de Catalunya, Barcelona.

Mateu-Figueras, G. and Barceló-Vidal, C.E. (eds) (2005) *Compositional Data Analysis Workshop – CoDaWork'05, Proceedings*, Universitat de Girona, Girona (E), ISBN: 84-8458-222-1, dugi-doc.udg.edu/handle/10256/617.

Mateu-Figueras, G. and Pawlowsky-Glahn, V. (2007) The skew-normal distribution on the simplex. *Communications in Statistics – Theory and Methods*, **36** (9), 1787–1802.

Mateu-Figueras, G., Pawlowsky-Glahn, V., and Egozcue, J.J. (2011) *The Principle of Working on Coordinates*, John Wiley & Sons, Ltd, pp. 31–42, 378 p.

Mateu-Figueras, G., Pawlowsky-Glahn, V., and Egozcue, J.J. (2013) The normal distribution in some constrained sample spaces. *SORT – Statistics and Operations Research Transactions*, **37** (1), 29–56.

McAlister, D. (1879) The law of the geometric mean. *Proceedings of the Royal Society of London*, **29**, 367–376.

Mert, C., Filzmoser, P., and Hron, K. (2014) Sparse principal balances. *Statistical Modelling*, 173–176.

Monti, G., Mateu-Figueras, G., and Pawlowsky-Glahn, V. (2011) *Notes on the Scaled Dirichlet Distribution*, John Wiley & Sons, Ltd, pp. 128–138. 378 p.

Mood, A.M., Graybill, F.A., and Boes, D.C. (1986) *Introduction to the Theory of Statistics*, 17th edn, McGraw-Hill, New York, 564 p.

Mosimann, J.E. (1962) On the compound multinomial distribution, the multivariate β-distribution and correlations among proportions. *Biometrika*, **49** (12), 65–82.

Nadarajah, S. and Kotz, S. (2007) Proportions, sums and ratios. *American Statistical Society*, **91**, 93–106.

Narayanan, A. (1991) Small sample properties of parameter estimation in the Dirichlet distribution. *Communications in Statistics B, Simulation and Computation*, **20** (2/3), 647–666.

Odum, E.P. (1971) *Fundamentals of Ecology*, Saunders, Philadelphia, PA.

Otero, N., Tolosana-Delgado, R., Soler, A., Pawlowsky-Glahn, V., and Canals, A. (2005) Relative vs. absolute statistical analysis of compositions: a comparative study of surface waters of a Mediterranean river. *Water Research*, **39** (7), 1404–1414.

Palarea-Albaladejo, J. and Martín-Fernández, J.A. (2008) A modified EM alr-algorithm for replacing rounded zeros in compositional data sets. *Computers and Geosciences*, **34** (8), 2233–2251.

Palarea-Albaladejo, J., Martín-Fernández, J.A., and Gómez-García, J.A. (2007) Parametric approach for dealing with compositional rounded zeros. *Mathematical Geology*, **39** (7), 625–645.

Pardo-Igúzquiza, E. and Heredia, J. (2011) *Spectral Analysis of Compositional Data in Cyclostratigraphy*, John Wiley & Sons, Ltd, pp. 282–289, 378 p.

Pawlowsky, V. (1984) On spurious spatial covariance between variables of constant sum. *Science de la Terre, Séries Informatique*, **21**, 107–113.

Pawlowsky, V. (1986) Räumliche Strukturanalyse und Schätzung ortsabhängiger Kompositionen mit Anwendungsbeispielen aus der Geologie. Ph.D. thesis, Fachbereich Geowissenschaften, Freie Universität Berlin, Berlin (D), 170 p.

Pawlowsky-Glahn, V. (2003) *Statistical Modelling on Coordinates*, Universitat de Girona, ISBN: 84-8458-111-X, http://ima.udg.es/Activitats/CoDaWork2003/.

Pawlowsky-Glahn, V. and Buccianti, A. (eds) (2011) *Compositional Data Analysis: Theory and Applications*, John Wiley & Sons, Ltd, 378 p.

Pawlowsky-Glahn, V. and Egozcue, J.J. (2001) Geometric approach to statistical analysis on the simplex. *Stochastic Environmental Research and Risk Assessment (SERRA)*, **15** (5), 384–398.

Pawlowsky-Glahn, V. and Egozcue, J.J. (2002) BLU estimators and compositional data. *Mathematical Geology*, **34** (3), 259–274.

Pawlowsky-Glahn, V. and Egozcue, J.J. (2006) *Compositional Data and Their Analysis: An Introduction*, Special Publications, vol. **264**, Geological Society, London, 212 p.

Pawlowsky-Glahn, V. and Egozcue, J.J. (2011) Exploring compositional data with the coda-dendrogram. *Austrian Journal of Statistics*, **40** (1 & 2), 103–113.

Pawlowsky-Glahn, V., Egozcue, J.J., and Tolosana-Delgado, R. (2011) *Principal Balances*, CIMNE, Barcelona, ISBN: 978-84-87867-76-7.

Pawlowsky-Glahn, V. and Olea, R.A. (2004) *Geostatistical Analysis of Compositional Data*, Number 7 in Studies in Mathematical Geology (ed. J.A. DeGraffenreid), Oxford University Press.

Pawlowsky-Glahn, V., Egozcue, J.J., and Lovell, D. (2013) The product space T (tools for compositional data with a total). In *Proceedings of the 5th International Workshop on Compositional Data Analysis* (eds K. Hron, P. Filzmoser, and M. Templ), pp. 143–152, ISBN: 978-3-200-03103-6, http://www.statistik.tuwien.ac.at/CoDaWork/CoDaWork2013 Proceedings.pdf.

Pearson, K. (1897) Mathematical contributions to the theory of evolution. On a form of spurious correlation which may arise when indices are used in the measurement of organs. *Proceedings of the Royal Society of London*, **LX**, 489–502.

Peña, D. (1991) *Estadística – Modelos y Métodos. I. Fundamentos y II. Modelos lineales y series temporales*, Alianza Editorial, Madrid (E), 1093 p.

Peña, D. (2002) *Análisis de datos multivariantes*, McGraw-Hill, Madrid (E), 539 p.

Pontriaguin, L.S. (1969) *Équations Différentielles Ordinaires*, MIR, Moscow, 399 pp.

Rao, C.R. (1973) *Linear Statistical Inference and Its Applications*, John Wiley & Sons, Inc., 625 p.

Richter, D.H. and Moore, J.G. (1966) Petrology of the Kilauea Iki lava lake, Hawaii. U.S. Geological Survey Professional Paper 537-B, B1-B26, cited in (1995).

Rizzo, M.L. and Székely, G.J. (2008) Energy: E-statistics (energy statistics). R-project. R package version 1.1-0.

Robert, C.P. (1994) *The Bayesian Choice. A Decision-Theoretic Motivation*, Springer-Verlag, New York, 436 p.

Rohatgi, V.K. (1976) *An Introduction to Probability Theory and Mathematical Statistics*, Wiley Series in Probability and Statistics, John Wiley & Sons, Inc., New York, 684 p.

Rollinson, H.R. (1995) *Using Geochemical Data: Evaluation, Presentation, Interpretation*, Longman Geochemistry Series, Longman Group Ltd., Essex, 352 p.

Sarmanov, O.V. and Vistelius, A.B. (1959) On the correlation of percentage values. *Doklady of the Academy of Sciences of the USSR – Earth Sciences Section*, **126**, 22–25.

Shao, J. (1998) *Mathematical Statistics*, Springer Texts in Statistics, Springer.

Simpson, E.H. (1951) The interpretation of interaction in contingency tables. *Journal of the Royal Statistical Society, Series B*, **13**, 238–241.

Smith, T. and Brunsdon, T. (1989) *The Time Series Analysis of Compositional Data*, American Statistical Association.

Solano-Acosta, W. and Dutta, P.K. (2005) Unexpected trend in the compositional maturity of second-cycle sand. *Sedimentary Geology*, **178** (3-4), 275–283.

Stevens, N.P., Bray, E.E., and Evans, E.D. (1956) Hydrocarbons in sediments of Gulf of Mexico. *American Association of Petroleum Geologists Bulletin*, **40** (5), 975–983.

Székely, G.J. and Rizzo, M.L. (2005) A new test for multivariate normality. *Journal of Multivariate Analysis*, **93** (1), 58–80.

Templ, M., Hron, K., and Filzmoser, P. (2010) robCompositions: Robust Estimation for Compositional Data. Manual and package, version 1.4.1. R-project.

Templ, M., Hron, K., and Filzmoser, P. (2011) *robCompositions: An R-package for Robust Statistical Analysis of Compositional Data*, John Wiley & Sons, Ltd, pp. 341–355, 378.

Thió-Henestrosa, Tolosana-Delgado, S.R., and Gómez, O. (2005) New features of CoDaPack–a compositional data package, in *Proceedings of IAMG'05 – The X Annual Conference of the Internationl Association for Mathematical Geology*, vol. **2** (eds Q. Cheng and G. Bonham-Carter), Association for Mathematical Geology, pp. 1171–1178, 1345 p., ISBN: 0-9734220-1-7.

Thió-Henestrosa, S. and Martín-Fernández, J.A. (eds) (2003) *Compositional Data Analysis Workshop – CoDaWork'03, Proceedings*, Universitat de Girona, ISBN: 84-8458-111-X, dugi-doc.udg.edu/handle/10256/616.

Thió-Henestrosa, S. and Martín-Fernández, J.A. (2005) Dealing with compositional data: the freeware CoDaPack. *Mathematical Geology*, **37** (7), 773–793.

Tolosana-Delgado, R. (2006) Geostatistics for constrained variables: positive data, compositions and probabilities. Application to environmental hazard monitoring. Ph.D. thesis, Universitat de Girona (Spain), 198 p.

Tolosana-Delgado, R. (2011) Guía para el análisis espacial de datos composicionales. guide for the spatial analysis of compositional data. *Boletín Geológico y Minero*, **122** (4), 469–482.

Tolosana-Delgado, R. and van den Boogaart, K.G. (2013a) Joint consistent mapping of high-dimensional geochemical surveys. *Mathematical Geosciences*, **45** (8), 983–1004.

Tolosana-Delgado, R. and van den Boogaart, K.G. (2013b) *Regression Between Compositional Data Sets*, Technische Universität Wien, pp. 163–176.

Tolosana-Delgado, R., van den Boogaart, K.G., Mikes, T., and Eynatten, H.V. (2008) *Statistical Treatment of Grain-Size Curves and Empirical Distributions: Densities as*

Compositions?, Universitat de Girona, http://hdl.handle.net/10256/723, http://ima.udg.es
/Activitats/CoDaWork2008/.

Tolosana-Delgado, R., van den Boogaart, K.G., and Pawlowsky-Glahn, V. (2009) Estimating
and modeling variograms of compositional data with occasional missing variables in R. In
StatGIS'09, Geoinformatics for Environmental Surveillance Workshop, Milos (Greece).

Tolosana-Delgado, R., van den Boogaart, K.G., and Pawlowsky-Glahn, V. (2011) *Geostatistics for Compositions*, John Wiley & Sons, Ltd, pp. 73–86. 378 p.

Tolosana-Delgado, R., Egozcue, J.J., Sánchez-Arcilla, A., and Gómez, J. (2011) Wave
height data assimilation using non-stationary kriging. *Computers and Geosciences*, **37**,
363–370.

Tolosana-Delgado, R. and Eynatten, H.V. (2009) Grain-size control on petrographic composition of sediments: compositional regression and rounded zeroes. *Mathematical Geosciences*, **41**, 869–886.

Tolosana-Delgado, R., Eynatten, H.V., and Karius, V. (2011) Constructing modal mineralogy from geochemical composition: a geometric-Bayesian approach. *Computers and
Geosciences*, **37** (5, SI), 677–691.

Tolosana-Delgado, R., Otero, N., Pawlowsky-Glahn, V., and Soler, A. (2005) Latent compositional factors in the llobregat river basin (Spain) hydrogeoeochemistry. *Mathematical
Geology*, **37** (7), 681–702.

Tolosana-Delgado, R., Pawlowsky-Glahn, V., and Egozcue, J.J. (2008) Indicator kriging
without order relation violations. *Mathematical Geosciences*, **40**, 327–347.

Verhulst, P.F. (1838) Notice sur la loi que la population poursuit dans son accroyssement.
Correspondance Mathématique et Fysique, **10**, 113–121.

Volterra, V. (1926) Variazioni e fluttuazioni del numero d'individui in specie animali con-
viventi. *Memoria della Regia Accademia Nazionale dei Lincei, Series*, **2**, 31–113.

Wikipedia (2014) Ergebnisse der Bundestagswahlen (results of elections to the german federal parliament). Wikipedia page.

Witting, H. (1985) *Mathematische Statistik I. Parametrische Verfahren bei festem Stich-
probenumpfang*, B. G. Teubner, Stuttgart (DE), 538 p.

Zee Ma, Y. (2009) Simpson's paradox in natural resource evaluation. *Mathematical Geosciences*, **41**, 193–213.

Appendix A

Practical recipes

A.1 Plotting a ternary diagram

Denote the three vertices of the ternary diagram counterclockwise from the upper vertex as A, B, and C (see Figure A.1). The scale of the plot is arbitrary and a unitary equilateral triangle can be chosen. Assume that $[u_0, v_0]$ are the plotting coordinates of the B vertex. The C vertex is then $C = [u_0 + 1, v_0]$ and the vertex A has abscissa $u_0 + 0.5$; the square-height is obtained using Pythagoras' theorem: $1^2 - 0.5^2 = 3/4$. Thus, $A = [u_0 + 0.5, v_0 + \sqrt{3}/2]$. These are the vertices of the triangle shown in Figure A.1, where the origin has been shifted to $[u_0, v_0]$ to center the plot. The figure is obtained by plotting the segments AB, BC, and CA.

To plot a sample point $\mathbf{x} = [x_1, x_2, x_3]$ in barycentric coordinates, closed to a constant κ, the corresponding plotting coordinates $[u, v]$ are needed. They are obtained as a convex linear combination of the plotting coordinates of the vertices

$$[u, v] = \frac{1}{\kappa}(x_1 A + x_2 B + x_3 C),$$

with

$$A = [u_0 + 0.5, v_0 + \sqrt{3}/2], \quad B = [u_0, v_0], \quad C = [u_0 + 1, v_0]. \tag{A.1}$$

Note that the coefficients of the convex linear combination must be closed to 1, which is obtained by dividing by κ.

Modeling and Analysis of Compositional Data, First Edition.
Vera Pawlowsky-Glahn, Juan José Egozcue and Raimon Tolosana-Delgado.
© 2015 John Wiley & Sons, Ltd. Published 2015 by John Wiley & Sons, Ltd.

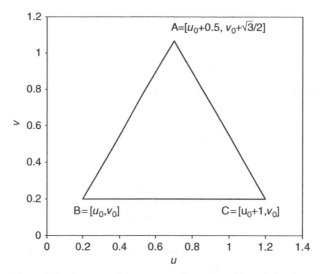

Figure A.1   Plot of the frame of a ternary diagram. The shift plotting coordinates are $[u_0, v_0] = [0.2, 0.2]$, and the length of each side of the triangle is 1.

To plot an Aitchison straight line in the ternary diagram, one can discretize it in a series of points and approximate it by a piecewise linear curve joining these points. Experience tells that it is more than enough to use around 100 points placed between −5 and +5 score of an ilr coordinate. Thus, to plot a line along the leading composition $\mathbf{b} = [b_1, b_2, b_3]$ passing through the compositional point $\mathbf{a} = [a_1, a_2, a_3]$, the following steps can be followed:

1. normalize the leading composition $\mathbf{b}$ to

$$\mathbf{w} = \frac{1}{\|\mathbf{b}\|_a} \odot \mathbf{b};$$

2. determine the composition $\mathbf{x}(0) = [x_1(0), x_2(0), x_3(0)]$ on the line that is nearest to the barycenter of the ternary diagram; due to standard properties of orthogonal projections, this is

$$\mathbf{x}(0) = \mathbf{a} \ominus \langle \mathbf{a}, \mathbf{w} \rangle_a \odot \mathbf{w};$$

3. take a parameter t between −5 and +5 in steps of 0.1 to calculate 101 points $\mathbf{x}(t) = \mathbf{x}(0) \oplus t \odot \mathbf{w};$

4. for each point $\mathbf{x}(t) = [x_1(t), x_2(t), x_3(t)]$, calculate a linear combination of the vertices of Equation (A.1)

$$[u(t), v(t)] = x_1(t) \cdot A + x_2(t) \cdot B + x_3(t) \cdot C \qquad \text{(A.2)}$$

and join all consecutive points with a segment.

A.2   Parameterization of an elliptic region

To plot an ellipse in $\mathbb{R}^2$ and to plot its backtransform in the ternary diagram, the plotting program needs a sequence of points that can be joined by a smooth curve. This requires the points to be in a certain order, so that they can be joint consecutively. It can be achieved by using polar coordinates, as they allow to give a consecutive sequence of angles that follow the border of the ellipse in one direction. The degree of approximation of the ellipse will depend on the number of points used for discretization.

The algorithm is based on the following reasoning. Imagine an ellipse located in $\mathbb{R}^2$, in which principal axes are not parallel to the axes of the Cartesian coordinate system. To express the points along the ellipse $\mathbf{x}^* = [x_1^*, x_2^*]$ in polar coordinates, we have to

1. translate the ellipse to the origin;

2. rotate it in such a way that the principal axis of the ellipse coincide with the axis of the coordinate system;

3. stretch the axis corresponding to the shorter principal axis in such a way that the ellipse becomes a circle in the new coordinate system;

4. this circle is defined in polar coordinates by a constant radius r equal to half the largest axis of the ellipse and an angle covering all values within the interval $(0, 2\pi)$ in radians; in practice, it is common to discretize the angle in 361 steps, including both extremes 0 and 2π;

5. transform the circle from polar coordinates to the original Cartesian coordinates using the expressions $x_1^* = r \cos \rho$, $x_2^* = r \sin \rho$;

6. undo steps 3 to 1 (in reverse order) to obtain the expression of the ellipse in terms of the polar coordinates.

Although this might sound tedious and complicated, some results from matrix theory reduce this procedure to some linear algebra and a problem of eigenvalues and eigenvectors. In fact, any symmetric matrix can be decomposed into the matrix product $\mathbf{Q}\boldsymbol{\Lambda}\mathbf{Q}^{\mathsf{T}}$, where $\boldsymbol{\Lambda}$ is the diagonal matrix of eigenvalues and $\mathbf{Q}$ is the matrix of orthonormal eigenvectors associated with them. For $\mathbf{Q}$, it holds that $\mathbf{Q}^{\mathsf{T}} = \mathbf{Q}^{-1}$, and therefore, $\mathbf{Q}^{-\mathsf{T}} = \mathbf{Q}$.

In general, interest lies in ellipses whose matrix is related to the sample covariance matrix $\hat{\boldsymbol{\Sigma}}$, particularly in its inverse. Substituting $\hat{\boldsymbol{\Sigma}}^{-1} = \mathbf{Q}\boldsymbol{\Lambda}^{-1}\mathbf{Q}^{\mathsf{T}}$ in the equation of the ellipse (7.7) or (7.8),

$$(\mathbf{x}^* - \boldsymbol{\mu})\mathbf{Q}\boldsymbol{\Lambda}^{-1}\mathbf{Q}^{\mathsf{T}}(\mathbf{x}^* - \boldsymbol{\mu})^{\mathsf{T}} = (\mathbf{Q}^{\mathsf{T}}(\mathbf{x}^* - \boldsymbol{\mu})^{\mathsf{T}})^{\mathsf{T}}\boldsymbol{\Lambda}^{-1}(\mathbf{Q}^{\mathsf{T}}(\mathbf{x}^* - \boldsymbol{\mu})^{\mathsf{T}}) = \kappa,$$

where $\mathbf{x}^*$ is a point along the ellipse and $\boldsymbol{\mu}$ describes its center, and the vectors are taken as rows as a default. The vector $\mathbf{Q}^{\mathsf{T}}(\mathbf{x}^* - \boldsymbol{\mu})^{\mathsf{T}}$ corresponds to a rotation in real space, where the new coordinate axis are precisely the eigenvectors. Given that $\boldsymbol{\Lambda}$ is a diagonal matrix, the next step consists of writing $\boldsymbol{\Lambda}^{-1} = \boldsymbol{\Lambda}^{-1/2}\boldsymbol{\Lambda}^{-1/2}$,

$$(\mathbf{Q}^{\mathsf{T}}(\mathbf{x}^* - \boldsymbol{\mu})^{\mathsf{T}})^{\mathsf{T}}\boldsymbol{\Lambda}^{-1/2}\boldsymbol{\Lambda}^{-1/2}(\mathbf{Q}^{\mathsf{T}}(\mathbf{x}^* - \boldsymbol{\mu})^{\mathsf{T}})$$
$$= (\boldsymbol{\Lambda}^{-1/2}\mathbf{Q}^{\mathsf{T}}(\mathbf{x}^* - \boldsymbol{\mu})^{\mathsf{T}})^{\mathsf{T}}(\boldsymbol{\Lambda}^{-1/2}\mathbf{Q}^{\mathsf{T}}(\mathbf{x}^* - \boldsymbol{\mu})^{\mathsf{T}}) = \kappa.$$

This transformation is equivalent to a rescaling of the basis vectors in such a way, that the ellipse becomes a circle of radius $\sqrt{\kappa}$, which is easy to express in polar coordinates,

$$\boldsymbol{\Lambda}^{-1/2}\mathbf{Q}^{\mathsf{T}}(\mathbf{x}^* - \boldsymbol{\mu})^{\mathsf{T}} = \begin{pmatrix} \sqrt{\kappa}\cos\theta \\ \sqrt{\kappa}\sin\theta \end{pmatrix}, \quad \text{or} \quad (\mathbf{x}^* - \boldsymbol{\mu})^{\mathsf{T}} = \mathbf{Q}\boldsymbol{\Lambda}^{1/2}\begin{pmatrix} \sqrt{\kappa}\cos\theta \\ \sqrt{\kappa}\sin\theta \end{pmatrix}.$$

The parameterization that we are looking for is thus given by

$$\boldsymbol{\mu}^{\mathsf{T}} = (\mathbf{x}^*)^{\mathsf{T}} - \mathbf{Q}\boldsymbol{\Lambda}^{1/2}\begin{pmatrix} \sqrt{\kappa}\cos\theta \\ \sqrt{\kappa}\sin\theta \end{pmatrix}.$$

Note that $\mathbf{Q}\boldsymbol{\Lambda}^{1/2}$ is the upper triangular matrix of the Cholesky decomposition of $\hat{\boldsymbol{\Sigma}}$,

$$\hat{\boldsymbol{\Sigma}} = \mathbf{Q}\boldsymbol{\Lambda}^{1/2}\boldsymbol{\Lambda}^{1/2}\mathbf{Q}^{\mathsf{T}} = (\mathbf{Q}\boldsymbol{\Lambda}^{1/2})(\boldsymbol{\Lambda}^{1/2}\mathbf{Q}^{\mathsf{T}}) = \mathbf{U}\mathbf{L};$$

and, from $\hat{\boldsymbol{\Sigma}} = \mathbf{U}\mathbf{L}$ and $\mathbf{L} = \mathbf{U}^{\mathsf{T}}$, the condition results

$$\begin{pmatrix} u_{11} & u_{12} \\ 0 & u_{22} \end{pmatrix}\begin{pmatrix} u_{11} & 0 \\ u_{12} & u_{22} \end{pmatrix} = \begin{pmatrix} \hat{\Sigma}_{11} & \hat{\Sigma}_{12} \\ \hat{\Sigma}_{12} & \hat{\Sigma}_{22} \end{pmatrix},$$

which implies

$$u_{22} = \sqrt{\hat{\Sigma}_{22}},$$

$$u_{12} = \frac{\hat{\Sigma}_{12}}{\sqrt{\hat{\Sigma}_{22}}},$$

$$u_{11} = \sqrt{\frac{\hat{\Sigma}_{11}\hat{\Sigma}_{22} - \hat{\Sigma}_{12}^2}{\hat{\Sigma}_{22}}} = \sqrt{\frac{|\hat{\Sigma}|}{\hat{\Sigma}_{22}}},$$

and, for each component of the vector $\mathbf{x}^*$,

$$x_1^* = \mu_1 - \sqrt{\frac{|\hat{\Sigma}|}{\hat{\Sigma}_{22}}}\sqrt{\kappa}\cos\theta - \frac{\hat{\Sigma}_{12}}{\sqrt{\hat{\Sigma}_{22}}}\sqrt{\kappa}\sin\theta,$$

$$x_2^* = \mu_2 - \sqrt{\hat{\Sigma}_{22}}\sqrt{\kappa}\sin\theta. \tag{A.3}$$

The points describing the ellipse in the simplex are $\mathbf{x} = \text{ilr}^{-1}(\mathbf{x}^*)$ (see Section 4.4). Following the same procedure as in the preceding section, the resulting points $\mathbf{x}$ can be drawn in a ternary diagram as linear combinations of the triangle vertices, as in Equation (A.2).

The procedures described apply to the three cases studied in Section 7.4, just using the covariance matrix $\hat{\Sigma}$ rescaled by the appropriate κ. This is obtained from a different distribution depending on the kind of region that is being built: probability region (χ^2 distribution), confidence region (Fisher $\mathcal{F}$ distribution) or predictive region (Student–Siegel distribution).

A.3   Matrix expressions of change of representation

Consider a composition $\mathbf{x}$ and two orthonormal basis in $\mathcal{S}^D$ with associated contrast matrices $\boldsymbol{\Psi}_0$ and $\boldsymbol{\Psi}_1$, such that the corresponding ilr-transformations are $\mathbf{x}_i^* = \text{ilr}_i(\mathbf{x}) = \ln \mathbf{x} \cdot \boldsymbol{\Psi}_i^\mathsf{T}$, $i = 0, 1$. Consider as well the vector of clr-scores $\boldsymbol{\xi} = \text{clr}(\mathbf{x})$ and the alr coordinates $\boldsymbol{\zeta} = \text{alr}(\mathbf{x})$. Also define the matrix $\mathbf{H} = \mathbf{I}_D - (1/D)\mathbf{1}_D^\mathsf{T}\mathbf{1}_D$ and the rectangular matrices $\mathbf{F} = [\mathbf{I}_{D-1} : \mathbf{1}_{D-1}]$ and $\mathbf{G}^\mathsf{T} = [\mathbf{I}_{D-1} : \mathbf{0}_{D-1}] - (1/D)\mathbf{1}_{D-1}^\mathsf{T}\mathbf{1}_D$. With this notation, the following matrix expressions allow to relate the several logratio representations of any composition:

$$\xi = \mathbf{x}_1^* \cdot \boldsymbol{\Psi}_1 = \mathbf{x}_2^* \cdot \boldsymbol{\Psi}_2 = \ln \mathbf{x} \cdot \mathbf{H} = \boldsymbol{\zeta} \cdot \mathbf{G}^\mathsf{T},$$

$$\mathbf{x}_1^* = \xi \cdot \boldsymbol{\Psi}_1^\mathsf{T} = \mathbf{x}_2^* \cdot (\boldsymbol{\Psi}_2 \cdot \boldsymbol{\Psi}_1^\mathsf{T}) = \boldsymbol{\zeta} \cdot \mathbf{G}^\mathsf{T} \cdot \boldsymbol{\Psi}_1^\mathsf{T},$$

$$\boldsymbol{\zeta} = \xi \cdot \mathbf{F}^\mathsf{T} = \mathbf{x}_1^* \cdot \boldsymbol{\Psi}_1 \cdot \mathbf{F}^\mathsf{T}.$$

These expressions apply not only to any datum but also to many statistics (a mean, a regression slope or gradient, a linear discriminant direction), for both its sample and population versions. The relations between clr and ilr apply also to principal component vectors. Proofs of the relations between alr and clr vectors can be found in Aitchison (1986), while those involving ilr were given by Egozcue et al. (2003).

Consider the same matrices defined before, and denote the variation matrix $\mathbf{T} = [t_{jk}]$, the covariance matrix of the clr-transformed data by $\boldsymbol{\Sigma}_c = [\sigma_{jk}]$, and the covariance matrix of the ilr-transformed data, using the contrast matrix $\boldsymbol{\Psi}_i^\mathsf{T}$, $i = 0, 1$ by $\boldsymbol{\Sigma}_i$. With this notation, the several representations of the variability of a random composition are related as follows

$$\boldsymbol{\Sigma}_1 = -\frac{1}{2}\boldsymbol{\Psi}_1 \cdot \mathbf{T} \cdot \boldsymbol{\Psi}_1^\mathsf{T} = \boldsymbol{\Psi}_1 \cdot \boldsymbol{\Sigma}_c \cdot \boldsymbol{\Psi}_1^\mathsf{T} = (\boldsymbol{\Psi}_1 \cdot \boldsymbol{\Psi}_2^\mathsf{T}) \cdot \boldsymbol{\Sigma}_2 \cdot (\boldsymbol{\Psi}_2 \cdot \boldsymbol{\Psi}_1^\mathsf{T}),$$

$$\boldsymbol{\Sigma}_c = -\frac{1}{2}\mathbf{H} \cdot \mathbf{T} \cdot \mathbf{H} = \boldsymbol{\Psi}_1^\mathsf{T} \cdot \boldsymbol{\Sigma}_1 \cdot \boldsymbol{\Psi}_1 = \boldsymbol{\Psi}_2^\mathsf{T} \cdot \boldsymbol{\Sigma}_2 \cdot \boldsymbol{\Psi}_2.$$

Note that the clr/ilr-representations of endomorphisms of Section 4.9.2 satisfy the same relations as clr/ilr-covariances. No simple matrix expression exists to compute the variation matrix from the ilr or clr covariance matrix, but elementwise, it holds that

$$t_{ij} = \sigma_{jj} + \sigma_{kk} - 2\sigma_{jk},$$

relating it to the clr covariance matrix. These expressions hold for both sample and population versions of these variability statistics. The relations between clr-covariances and variation were given by Aitchison (1986). Expressions involving any basis, any ilr in particular, can be found in Tolosana-Delgado (2006).

Appendix B

Random variables

In a general sense, random variables are measurable functions from a probability space on a sample space. This sentence involves some mathematical concepts that are recalled here for easy reference when reading Chapter 6. Some are elementary, but others are linked to abstract measure theory. Most concepts can be easily found in textbooks on probability, for instance, Mood et al. (1986), Ash (1972), or Feller (1966). The first section of this appendix presents the definitions and first properties of probability spaces and random variables. The second section is a reference of the standard description of a probability measure using the cumulative distribution function and the probability density.

B.1 Probability spaces and random variables

A probability space is made of three mathematical objects, an arbitrary set, here denoted by Ω, a σ-field defined on Ω, and a probability measure. The set Ω has finite or infinite cardinal, that is, the number of its elements is finite or infinite.

To define a probability space, a σ-field must be defined in Ω.

Definition B.1 (σ-field).
Let Ω be an arbitrary set. A set $\mathcal{A}$ of subsets of Ω is a σ-field on Ω if it satisfies the following conditions:

1. The whole set Ω is in $\mathcal{A}$;

2. If $A \in \mathcal{A}$, and $\bar{A}$ is its complement in Ω, then $\bar{A} \in \mathcal{A}$;

Modeling and Analysis of Compositional Data, First Edition.
Vera Pawlowsky-Glahn, Juan José Egozcue and Raimon Tolosana-Delgado.
© 2015 John Wiley & Sons, Ltd. Published 2015 by John Wiley & Sons, Ltd.

3. *If A_1 and A_2 are subsets of Ω in $\mathcal{A}$, then $A_1 \cup A_2$ is also in $\mathcal{A}$;*

4. *For $A_1, A_2, \ldots$, a sequence of subsets of Ω in $\mathcal{A}$, it holds*

$$\bigcup_{i=1}^{\infty} A_i \in \mathcal{A}.$$

The first three conditions in this definition characterize what is known as a Boolean field. Condition 4 is a generalization of condition 3. It gives the name of σ-field to $\mathcal{A}$ and is a key property for the definition of probability on spaces Ω in which cardinal is infinite. Probability and any measure are only defined for subsets of Ω in the σ-field; they are commonly called *events*.

Some conclusions are immediate from Definition B.1. Among them, for example, the fact that the empty set is included in the σ-field, $\emptyset \in \mathcal{A}$; that when the cardinal of Ω is finite, the set of all subsets of Ω is a σ-field; or that any intersection (finite or infinite) of elements in $\mathcal{A}$ is also an element of Ω.

The pair $(\Omega, \mathcal{A})$ constitutes a measurable space. The reason for this name is that measures, specifically probability measures, need to be defined on measurable spaces.

Definition B.2 (σ-additive, positive measure).
Consider a measurable space $(\Omega, \mathcal{A})$. A function $\mu : \mathcal{A} \to \mathbb{R}_+$ is a sigma-additive positive measure on $\mathcal{A}$ if it satisfies the following conditions:

1. *(positivity) If $A \in \mathcal{A}$, then $\mu[A] \geq 0$;*

2. *(σ-additivity) Let be $\{A_1, A_2, \ldots\}$ a finite or infinite sequence of non-overlapping subsets of Ω in $\mathcal{A}$, that is, for $i \neq j$, $A_i \cap A_j = \emptyset$, then*

$$\mu\left[\bigcup_i A_i\right] = \sum_i \mu[A_i].$$

Definition B.3 (probability measure).
Consider a measurable space $(\Omega, \mathcal{A})$. A σ-additive positive measure P is a probability measure on $\mathcal{A}$ if $P[\Omega] = 1$.

Although traditionally $P[\Omega] = 1$, it is also common to use probabilities in percentage, meaning implicitly $P[\Omega] = 100$. In general, what is important is that a probability is a finite σ-additive positive measure. The value assigned to $P[\Omega]$ is a convention that facilitates computations and handling of probabilities, but it is not an intrinsic property. The fact that probabilities can be multiplied by an arbitrary positive constant, for instance 100, and that they preserve their meaning, matches the principle of scale invariance in compositional data analysis described in Section 2.2.1. See Egozcue et al. (2013) for a discussion.

Definition B.4 (probability space).
An arbitrary set Ω, with a σ-field $\mathcal{A}$ defined on it and a probability measure P defined on $\mathcal{A}$, is called probability space and is denoted by $(\Omega, \mathcal{A}, P)$.

The previous definitions can be considered as axioms for a probability theory. All elementary properties of probability can be derived using these. A general random variable can be then defined as follows.

Definition B.5 (random variable).
A random variable X is a function from a probability space $(\Omega, \mathcal{A}, P_\Omega)$ on a measurable space $(S, \mathcal{B})$ such that, for all $B \in \mathcal{B}$, $X^{-1}[B] \in \mathcal{A}$.

The condition that any element of the σ-field $\mathcal{B}$ (called event) has anti-image by X in $\mathcal{A}$, characterizes X as a measurable function. The random variable X allows to define a probability on $(S, \mathcal{B})$, so that it is also a probability space. The construction is quite simple and the main point is setting $P[B] = P_\Omega[X^{-1}[B]]$, for all events B in $\mathcal{B}$. Then, $(S, \mathcal{B}, P)$ is also a probability space which is called *sample space*.

The very weak structure of both probability spaces $(\Omega, \mathcal{A}, P_\Omega)$, $(S, \mathcal{B}, P)$, gives the opportunity of using them for a wide range of models. For instance, when S is a space of functions, the random variable is called stochastic process; if the elements S are sets, then X is a random set; for $S = \mathbb{R}$, X is a standard real random variable. When S is the D-part simplex, the random variable is a random composition.

As stated in Chapter 6, once a sample space is selected as a model for a random variable, some structures beyond the σ-field can be inherited. The case in which S is $\mathbb{R}$ provides the use of addition and distance in $\mathbb{R}$ for real random variables. Similarly, when the sample space is the simplex, the Aitchison geometry becomes available, thus allowing perturbation, powering, and the use of distances or projections of random compositions.

An important issue in probability theory is the use of conditional probabilities. They appear as the result of a change of the sample space.

Definition B.6 (conditional probability).
Let X be a random variable with sample space $(S, \mathcal{B}, P)$ and $S_1 \subseteq S$ an element of the σ-field $\mathcal{B}$ such that $P[S_1] \neq 0$. The probability of an event $B \in \mathcal{B}$, conditional to S_1, is

$$P[X \in B | S_1] = \frac{P[X \in (B \cap S_1)]}{P[S_1]}. \tag{B.1}$$

This definition states that the original sample space of X, $(S, \mathcal{B}, P)$, has been restricted to $(S_1, \mathcal{B}_1, P_1)$, where the elements of $\mathcal{B}_1$ are the intersections of those of $\mathcal{B}$ with S_1. The conditional probability is then identified with P_1, that is, $P_1[X \in$

$B] = P[X \in B|S_1]$, and the ratio to $P[S_1]$ in Equation (B.1) is just a normalization to force $P_1[S_1] = 1$.

Conditional probability is related to the most frequently used results in probability theory, the total probability theorem, and Bayes' formula. In their discrete versions, they can be formulated as follows.

Theorem B.7   (total probability).
Let $(S, \mathcal{B}, P)$ be a probability space and $B_1, B_2, \ldots$ a partition of S such that each $B_i \in \mathcal{B}$. Let A be an event in S, then

$$P[A] = \sum_i P[A|B_i] \cdot P[B_i].$$

Theorem B.8   (Bayes' formula).
Let $(S, \mathcal{B}, P)$ be a probability space and $B_1, B_2, \ldots$ a partition of S such that each $B_i \in \mathcal{B}$. Let A be an event in S, then

$$P[B_i|A] = \frac{1}{C}P[A|B_i] \cdot P[B_i], \quad C = P[A] = \sum_j P[A|B_j] \cdot P[B_j],$$

for $i = 1, 2, \ldots$

The importance of Bayes' formula, and its continuous counterpart, is that it is an information acquisition paradigm. Interest is in the occurrence of the events B_i. Initially, or a priori, there is uncertainty about the events. This uncertainty is quantified by $P[B_i]$, $i = 1, 2, \ldots$ An experiment is then conducted and a result A is obtained. The information provided by this result A is coded as the conditional probability $L(B_i|A) = P[A|B_i]$, known as the likelihood of B_i given the experimental result A. Bayes' formula explicitly shows that, up to a multiplicative constant $1/C$, the probability of the events B_i after the experimental result A is the product of the prior probabilities and the likelihood associated with the experimental result A. This motivated calling the $P[B_i|A]$, $i = 1, 2, \ldots$, *posterior probabilities*.

When dealing with a finite partition of S, $B_1, B_2, \ldots, B_D$, Bayes' formula (B.8) has a direct interpretation in the Aitchison geometry of the simplex. In fact, consider $p_1 = P[B_1], p_2 = P[B_2], \ldots, p_D = P[B_D], \sum_1^D p_i = 1$, grouped in a composition $\mathbf{p} \in S^D$. Also consider the composition $\mathbf{q} \in S^D$, in which parts are proportional to the likelihood, that is, $\mathbf{q} = C[L(B_i|A)]$. Then, the perturbation $\mathbf{q} \oplus \mathbf{p}$ equals the posterior probabilities. The factor $1/C$ in the Bayes' formula is just the closure in the perturbation. Also, taking a subcomposition is related to a case of conditional probability in Equation B.1. For instance, take $S_1 = B_1 \cup B_2 \cup B_3$, then

$$[P[B_1|S_1], P[B_2|S_1], P[B_3|S_1]] = C[P[B_1], P[B_2], P[B_3]],$$

that is, conditioning is reduced to a closure of a subcomposition.

B.2 Description of probability

The probability measure in the sample space of a random variable can be described in different ways depending on the characteristics of the sample space and the random variable. In the case of sample spaces with a numerable set of elements (elementary events), specification of probability of each elementary event is enough for determining any probability. When the sample space S has cardinal of the order of that of $\mathbb{R}$ or more, the description by probabilities of elementary events (the elements of S) becomes unavoidable, as infinitely many elementary events have null probability. The most used way of representing probabilities is based on considering an infinite family of elements of the σ-field parameterized by some real parameter. This family generates the whole σ-field, and the probabilities of the events in the family determine the probability of any event. In order to clarify this strategy, the standard example of a random variable with sample space $S = \mathbb{R}^n$ is sketched. The standard σ-field in $\mathbb{R}^n$ is that of the Borelian sets.

Definition B.9 (Borel σ-field).
Consider all open sets in $\mathbb{R}^n$. The minimal σ-field that contains all open sets of $\mathbb{R}^n$ is called Borel σ-field and is denoted by $\mathcal{B}(\mathbb{R}^n)$.

The Borel σ-field $\mathcal{B}(\mathbb{R}^n)$ has simple subsets of $\mathcal{B}(\mathbb{R}^n)$ as elements, such as individual points, intervals, orthants, and their combinations by union and intersection. All events needed in practice are included in $\mathcal{B}(\mathbb{R}^n)$ and their probability can be determined. However, there are infinitely many subsets of $\mathbb{R}^n$ not included in $\mathcal{B}(\mathbb{R}^n)$, although their construction is not elementary. See Kolmogorov and Fomin (1957) for an example.

Consider the family of all orthants with vertex at a point $\mathbf{x} = (x_1, x_2, \ldots, x_n)$ of $\mathbb{R}^n$, that is, the family of sets

$$R(\mathbf{x}) = \{(y_1, y_2, \ldots, y_n) \in \mathbb{R}^n \mid y_1 \leq x_1, y_2 \leq x_2, \ldots, y_n \leq x_n\},$$

and assume that $P[X \in R(\mathbf{x})]$ is specified for all $\mathbf{x} \in \mathbb{R}^n$. It should be noted that the definition of orthants depends on the particular basis of the space $\mathbb{R}^n$ that has been selected. It can be shown that the probability of all events in $\mathcal{B}(\mathbb{R}^n)$ is determined by $P[X \in R(\mathbf{x})]$ (Ash, 1972). The function $F(\mathbf{x}) = P[X \in R(\mathbf{x})]$ is called (cumulative) distribution function of the random variable X. The univariate case in $\mathbb{R}$ gives the familiar expression $F(x) = P[X \leq x]$.

The more popular and useful description of probability of a random variable is the probability density function. This description of the probability measure is only possible for the so-called (absolutely) continuous random variables. This means that the corresponding probability measure is absolutely continuous with respect to the reference measure considered in the sample space. Probability density functions, if they exist, do not depend on a particular choice of a basis of the space, but they depend on the reference measure defined on the space. For the real sample space, $\mathbb{R}^n$, the standard reference measure is the Lebesgue measure λ, while for the simplex, S^n, the standard reference measure is the Aitchison measure (Section 6.3).

Definition B.10 (absolutely continuous probability measure).

Let (S, B, P) be a sample space and $\lambda : B \to \mathbb{R}$ the reference measure in S. The probability measure P is absolutely continuous with respect to λ if, for any $B \in B$, $\lambda[B] = 0$ implies $P[B] = 0$.

Note that this definition corresponds to the idea that there is no accumulation of probability on individual points for which the reference measure is null. Consequently, the probability of the random variable taking value at an individual point is null.

Theorem B.11   (Radon–Nikodym derivative).

Let (S, B, P) be a sample space of the random variable X and $\lambda : B \to \mathbb{R}$ the reference measure in S. If P is an absolutely continuous probability measure with respect to the reference measure λ in S, then there exists a nonnegative function $f : S \to \mathbb{R}_+$ such that, for any $B \in B$,

$$P[X \in B] = \int_B f \, d\lambda, \tag{B.2}$$

where the function f is the Radon–Nikodym derivative of P with respect to λ. It is denoted by $f = dP/d\lambda$. In a context of probability theory, f is also called probability density function with respect to the reference measure λ.

When the sample space is $\mathbb{R}^n$ and the reference measure is the Lebesgue measure, Equation (B.2) becomes

$$P[X \in B] = \int_B f(\mathbf{x}) \, d\mathbf{x}.$$

As a summary, integrals of a probability density function on a set give the probability of the random variable being in that set. But the integral must be carried out with respect to the reference measure.

Author Index

General Index

Modeling and Analysis of Compositional Data, First Edition.
Vera Pawlowsky-Glahn, Juan José Egozcue and Raimon Tolosana-Delgado.
© 2015 John Wiley & Sons, Ltd. Published 2015 by John Wiley & Sons, Ltd.

List of abbreviations and symbols

Abbreviations and notation are introduced in the text when used for the first time. However, certain standards are adopted across the book or in some chapters. The following is a list of the main abbreviations and symbols, with the page in which they are first used or defined.

cdf	cumulative distribution function	105
pdf	probability density function	112
SBP	sequential binary partition	38
sv-function	simplex-valued function	183
$\mathbf{x}, \alpha$	row vectors, denoted in bold lower case	8
$\mathbf{A}, \mathbf{M}$	matrices, denoted in bold upper case	54
X, Y	random variables, in upper case	105
$X \sim$	distribution of X	114
$\mathcal{C}\mathbf{x}$	closure of the vector $\mathbf{x}$ (positive components)	9
alr	additive logratio	46
clr	centred logratio	35
ilr	isometric logratio	37
D	number of parts of a composition	8
$\mathbf{n}$	neutral element of S^D	25
S^D	D-part simplex	10
$\mathbb{R}^k$	k-dimensional real or Euclidean space	28
$\mathbb{R}^k_+$	k-dimensional positive orthant of $\mathbb{R}^k$	9
$[x_1, x_2, \dots, x_D]$	row vector of D components	8
$(\text{matrix})^\top$	transposition of matrices	36

Modeling and Analysis of Compositional Data, First Edition.
Vera Pawlowsky-Glahn, Juan José Egozcue and Raimon Tolosana-Delgado.
© 2015 John Wiley & Sons, Ltd. Published 2015 by John Wiley & Sons, Ltd.

$\oplus$	perturbation, group operation in the simplex	24
$\ominus$	perturbation-difference	25
$\odot$	powering, multiplication by scalars in the simplex	24
$\odot$ (matrix)	simplicial matrix product of a real vector and a compositional matrix	54
$\boxdot$ (matrix)	simplicial matrix product of a vector in S^D and a matrix of scalars	56
$\boldsymbol{\Psi}$	contrast matrix of a basis of the simplex	36
$\mathbf{I}_k$	(k, k)-identity matrix	36
$\mathbf{1}_k$	k-dimensional row vector with unitary components	36
$\mathbf{x}^*, \mathbf{y}^*$	orthonormal coordinates of the simplex	36
$g_m(\cdot)$	geometric mean of its arguments	34
$\hat{\mathbf{g}}$	row vector of geometric means across a compositional sample	66
$\partial, \partial_\oplus$	derivatives, ordinary and simplicial	183
$\langle \cdot, \cdot \rangle$	inner product in $\mathbb{R}^k$	35
$\langle \cdot, \cdot \rangle_a$	inner product in S^D	26
$d(\cdot, \cdot)$	distance in $\mathbb{R}^k$	35
$d_a(\cdot, \cdot)$	Aitchison distance in S^D	26
$\| \cdot \|$	norm in $\mathbb{R}^k$	35
$\| \cdot \|_a$	Aitchison norm in S^D	26
$P[\cdot]$	probability	105
$E[\cdot]$	expectation	109
n	size of a sample of data	65
cen$[\cdot]$	center	66
var$[\cdot]$	variance	109
cov$[\cdot]$	covariance	109
$\hat{(\cdot)}, \overline{(\cdot)}$	estimator, sample value; average, sample mean	109
totvar$[\cdot]$	total variance	67
$\Gamma(\cdot)$	Euler Gamma function	121
I{condition}	set indicator function	121

Statistics in Practice

Human and Biological Sciences

Berger – Selection Bias and Covariate Imbalances in Randomized Clinical Trials

Berger and Wong – An Introduction to Optimal Designs for Social and Biomedical Research

Brown, Gregory, Twelves and Brown – A Practical Guide to Designing Phase II Trials in Oncology

Brown and Prescott – Applied Mixed Models in Medicine, Third Edition

Campbell and Walters – How to Design, Analyse and Report Cluster Randomised Trials in Medicine and Health Related Research

Carpenter and Kenward – Multiple Imputation and its Application

Carstensen – Comparing Clinical Measurement Methods

Chevret (Ed.) – Statistical Methods for Dose-Finding Experiments

Cooke – Uncertainty Modeling in Dose Response: Bench Testing Environmental Toxicity

Eldridge – A Practical Guide to Cluster Randomised Trials in Health Services Research

Ellenberg, Fleming and DeMets – Data Monitoring Committees in Clinical Trials: A Practical Perspective

Gould (Ed) – Statistical Methods for Evaluating Safety in Medical Product Development

Hauschke, Steinijans and Pigeot – Bioequivalence Studies in Drug Development: Methods and Applications

Whitehead – Design and Analysis of Sequential Clinical Trials, Revised Second Edition

Whitehead – Meta-Analysis of Controlled Clinical Trials

Willan and Briggs – Statistical Analysis of Cost Effectiveness Data

Winkel and Zhang – Statistical Development of Quality in Medicine

Zhou, Zhou, Lui and Ding – Applied Missing Data Analysis in the Health Sciences

Earth and Environmental Sciences

Buck, Cavanagh and Litton – Bayesian Approach to Interpreting Archaeological Data

Chandler and Scott – Statistical Methods for Trend Detection and Analysis in the Environmental Statistics

Christie, Cliffe, Dawid and Senn (Eds.) – Simplicity, Complexity and Modelling

Gibbons, Bhaumik and Aryal – Statistical Methods for Groundwater Monitoring, 2nd Edition

Haas – Improving Natural Resource Management: Ecological and Political Models

Haas – Introduction to Probability and Statistics for Ecosystem Managers

Helsel – Nondetects and Data Analysis: Statistics for Censored Environmental Data

Illian, Penttinen, Stoyan and Stoyan – Statistical Analysis and Modelling of Spatial Point Patterns

Mateu and Muller (Eds) – Spatio-Temporal Design: Advances in Efficient Data Acquisition

McBride – Using Statistical Methods for Water Quality Management

Ofungwu – Statistical Applications for Environmental Analysis and Risk Assessment

Okabe and Sugihara – Spatial Analysis Along Networks: Statistical and Computational Methods

Pawlowsky-Glahn, Egozcue and Tolosana-Delgado – Modeling and Analysis of Compositional Data

Webster and Oliver – Geostatistics for Environmental Scientists, Second Edition

Wymer (Ed.) – Statistical Framework for RecreationalWater Quality Criteria and Monitoring

Industry, Commerce and Finance

Aitken – Statistics and the Evaluation of Evidence for Forensic Scientists, Second Edition

Balding – Weight-of-evidence for Forensic DNA Profiles

Brandimarte – Numerical Methods in Finance and Economics: A MATLAB-Based Introduction, Second Edition

Brandimarte and Zotteri – Introduction to Distribution Logistics

Chan – Simulation Techniques in Financial Risk Management

Coleman, Greenfield, Stewardson and Montgomery (Eds) – Statistical Practice in Business and Industry

Frisen (Ed.) – Financial Surveillance

Fung and Hu – Statistical DNA Forensics

Gusti Ngurah Agung – Time Series Data Analysis Using EViews

Jank and Shmueli – Modeling Online Auctions

Jank and Shmueli (Ed.) – Statistical Methods in e-Commerce Research

Lloyd – Data Driven Business Decisions

Kenett (Ed.) – Operational Risk Management: A Practical Approach to Intelligent Data Analysis

Kenett (Ed.) – Modern Analysis of Customer Surveys: With Applications using R

Kenett and Zacks – Modern Industrial Statistics: With Applications in R, MINITAB and JMP, Second Edition

Kruger and Xie – Statistical Monitoring of Complex Multivariate Processes: With Applications in Industrial Process Control

Lehtonen and Pahkinen – Practical Methods for Design and Analysis of Complex Surveys, Second Edition

Mallick, Gold, and Baladandayuthapani – Bayesian Analysis of Gene Expression Data

Ohser and Mücklich – Statistical Analysis of Microstructures in Materials Science

Pasiouras (Ed.) – Efficiency and Productivity Growth: Modelling in the Financial Services Industry

Pawlowsky-Glahn, Egozcue and Tolosana-Delgado – Modeling and Analysis of Compositional Data

Pfaff – Financial Risk Modelling and Portfolio Optimization with R

Pourret, Naim and Marcot (Eds) – Bayesian Networks: A Practical Guide to Applications

Rausand – Risk Assessment: Theory, Methods, and Applications

Ruggeri, Kenett and Faltin – Encyclopedia of Statistics and Reliability

Taroni, Biedermann, Bozza, Garbolino and Aitken – Bayesian Networks for Probabilistic Inference and Decision Analysis in Forensic Science, Second Edition

Taroni, Bozza, Biedermann, Garbolino and Aitken – Data Analysis in Forensic Science

Printed and bound by CPI Group (UK) Ltd, Croydon, CR0 4YY

27/10/2024

14580208-0001